A call to action is implicit in the rich legacy of King Bhumibol Adulyadej's 70-year reign. The call resounds in His Majesty's "Sufficiency Economy Philosophy", at the heart of his legacy for us in Thailand and the entire world. Indeed, Sufficiency principles have formed the core of Thailand's contribution to the United Nations General Assembly during its 2016 chairmanship of G-77, the group of 134 developing countries. During this period, G-77 helped launch the global campaign to achieve 17 Sustainable Development Goals by 2030, with Sufficiency Economy as one of the models for action that countries can adapt for their own development activities in achieving the SDGs.

This past year, as one of its contributions to the sustainability campaign, the Crown Property Bureau sponsored publication of *Thailand's Sustainable Development Sourcebook*. That volume has been well received as an information resource on the status of our country's preparedness. This year we have helped produce this second edition, titled *A Call to Action: Thailand and the Sustainable Development Goals*, as a concise accounting of how those resources can be directed in the national effort. The concise format should appeal to professionals, students and others working on the front lines of sustainability, in following His Majesty's footsteps.

The Crown Property Bureau, while dedicated to the prudent management of Crown assets, also seeks to improve the quality of life for all stakeholders in both public and private sectors. We apply Sufficiency Economy principles in the social and economic development activities of our four foundations: the Thailand Sustainable Development Foundation (TSDF), the Foundation of Virtuous Youth (FVY), the CPB Foundation, and Bookkhon Porpeang Foundation. A fifth CPB investment for a sound future is our support for the planned King Bhumibol Adulyadej Institute of Ethics and Education for Sustainable Development. KIEES will help lead the growing movement to apply a "sufficiency mindset" in fostering the sustainability of national life.

For 70 years King Bhumibol Adulyadej carried the torch of Thailand's sustainable development. On the 13th of October 2016 His Majesty passed the torch to us all, to continue meeting the challenges in the world's development scenario. We hope that readers of this volume will learn from His Majesty's legacy and apply its wisdom in advancing the global agenda of sustainability.

Chirayu Isarangkun Na Ayuthaya
Director-General
The Crown Property Bureau

A CALL TO ACTION

THAILAND AND THE SUSTAINABLE DEVELOPMENT GOALS

A CALL TO ACTION

THAILAND AND THE SUSTAINABLE DEVELOPMENT GOALS

edm

Editors
Will Baxter
Nicholas Grossman
Nina Wegner

Special Advisor
Priyanut Dharmapiya

Assistant Editor
Sutawan Chanprasert

Contributing Editors
Orawan Yafa
Pasamon Siriboon

Art Directors
Luxana Kiratibhongse
Benjapa Sodsathit
Patinya Rojnukkarin

Designers
Siree Simaraks
Nichapat Chokchaingamsangah
Chanthipapha Sopanaphimon

Project Coordinator
Jaruphan Phan-in

Production Manager
Annie Teo

Page 2: An estimated 490 acres of farmland have been irrigated by the Huay Klai Reservoir project in the northeast of Thailand.

Opposite page: The Chao Phraya River in Bangkok.

For a list of all writers and contributors, see pages 178-179.

First published in 2016 by
Editions Didier Millet (EDM)

Email: edm@edmbooks.com.sg
www.edmbooks.com

Bangkok office
Room 1310, 3rd Floor
8, Sukhumvit 49/9, Klongton Nua
Wattana, Bangkok 10110
Thailand
Tel: +66-2018-7808

Singapore head office
35B Boat Quay
Singapore 049824
Tel: +65-6324-9260

Cover design by
Palotai Design Co., Ltd.
www.palotaidesign.com

A special thank you to

for their support of this publication.

Color separation by
PICA DIGITAL, Singapore

Printed by
TWP SDN BHD, Malaysia

All rights reserved. No part of this publication may be reproduced or transmitted in any form or by any means, electronic or mechanical, or by any information storage and retrieval system, without the prior written permission of the copyright owners or publisher.

ISBN 978-981-4610-42-1

TABLE OF CONTENTS

9 Preface

10 Maps of Thailand

12 Thailand Overview

14 Introduction

28 Thailand's SDG Index Performance

30	**Goal 1:** No Poverty	114	**Goal 11:** Sustainable Cities and Communities
38	**Goal 2:** Zero Hunger	124	**Goal 12:** Responsible Consumption and Production
46	**Goal 3:** Good Health and Well-being	132	**Goal 13:** Climate Action
56	**Goal 4:** Quality Education	140	**Goal 14:** Life Below Water
66	**Goal 5:** Gender Equality	150	**Goal 15:** Life on Land
76	**Goal 6:** Clean Water and Sanitation	160	**Goal 16:** Peace, Justice and Strong Institutions
84	**Goal 7:** Affordable and Clean Energy	170	**Goal 17:** Partnerships for the Goals
92	**Goal 8:** Decent Work and Economic Growth		
98	**Goal 9:** Industry, Innovation and Infrastructure	178	Editorial Credits
106	**Goal 10:** Reduced Inequalities	180	Directory
		183	Picture Credits

PREFACE

The 2030 Agenda for Sustainable Development is a call to action that seeks to redress some of the egregious imbalances that human activities have created. These imbalances appear in the Earth's atmosphere, where carbon dioxide levels have reached a level unknown for 300 million years; in our seas, where fish stocks have been depleted and natural ecosystems disrupted; in our forests – or what remains of them – and throughout our society, where inequalities are leading to social turmoil. As these pages discuss, these imbalances represent a complex threat to our future from many different angles.

The concept of sustainable development is the paradigm shift that offers a new path forward, one that better balances economic, environmental and social concerns in a more holistic, altruistic and conscientious manner. For sustainable development to be truly attained, this balance must be realized not only inside our homes and businesses and across government-managed sectors, but within ourselves. In Thailand, the concept of sustainable development has special resonance, owing to the work of His Majesty King Bhumibol Adulyadej. His thousands of royally initiated projects in Bangkok and upcountry, and his values-based development strategy known as the Sufficiency Economy Philosophy (SEP), bear the hallmarks and share many of the same principles of the sustainable development approach.

Thailand is widely considered one of the countries that has begun to take the sustainable development movement seriously, but – as with all countries – there is a lot of work to be done. This book examines Thailand's progress and its challenges through the frame of the most famous sustainable development platform today – the 17 Sustainable Development Goals (SDGs) of the 2030 Agenda for Sustainable Development that was officially launched by the United Nations in 2016. This ambitious global program, which Thailand has joined, sets out 169 targets for nations to strive toward.

Some of the content in these pages is based on *Thailand's Sustainable Development Sourcebook*, which was published in November 2015. *A Call to Action: Thailand and the Sustainable Development Goals* includes much new information and discussion as well as updated statistics. The hope is that the articles here will both inform readers and inspire sincere interest in SEP and the SDGs, which will form key components of the national and international dialogue for years to come. One must ask but only you can decide: If a new mindset based on the values of SEP and sustainable development were adopted, would our future prospects, on many levels, be improved?

ACKNOWLEDGEMENTS

This book relied on the contributions of many people and organizations, many of whom are acknowledged on pages 178–179. We are indebted to the work of the many local and international organizations for the reports they have produced, and also to our advisors, partners and sponsors for making such a wide-ranging publication possible. In particular, we appreciate the guidance provided by the Thailand Sustainable Development Foundation, which is focused on supporting initiatives (like this one) that have been developed in the name of creating sustainable development in Thailand. We would like to extend a special thank you to Dr Chirayu Isarangkun Na Ayuthaya, Dr Kasem Watanachai, Kitti Wasinondh, Ongorn Abhakorn Na Ayuthaya, Pattamaniti Senanarong, Molraudee Saratun, Kim Atkinson, Ar-tara Satraroj, Yvan Van Outrive and the Thailand Development Research Institute.

PROVINCES OF THAILAND

NORTH
1. Mae Hong Son
2. Chiang Mai
3. Chiang Rai
4. Phayao
5. Lampang
6. Lamphun
7. Phrae
8. Nan
9. Uttaradit

WEST
10. Tak
11. Kanchanaburi
12. Ratchaburi
13. Phetchaburi
14. Prachuap Khiri Khan

CENTRAL
15. Sukhothai
16. Phitsanulok
17. Phetchabun
18. Kamphaeng Phet
19. Phichit
20. Nakhon Sawan
21. Uthai Thani
22. Chainat
23. Singburi
24. Lopburi
25. Suphanburi
26. Ang Thong
27. Saraburi
28. Ayudhya
29. Nakhon Nayok
30. Nakhon Pathom
31. Pathum Thani
32. Nonthaburi
33. Samut Songkhram
34. Samut Sakhon
35. **Bangkok**
36. Samut Prakan

NORTHEAST (Isan)
37. Loei
38. Nong Khai
39. Nong Bua Lamphu
40. Udon Thani
41. Sakon Nakhon
42. Nakhon Phanom
43. Khon Kaen
44. Kalasin
45. Mukdahan
46. Chaiyaphum
47. Maha Sarakham
48. Roi Et
49. Yasothon
50. Amnat Charoen
51. Nakhon Ratchasima (Korat)
52. Buriram
53. Surin
54. Si Saket
55. Ubon Ratchathani
56. Bueng Kan

EAST
57. Prachinburi
58. Chachoengsao
59. Sa Kaeo
60. Chonburi
61. Rayong
62. Chanthaburi
63. Trat

SOUTH
64. Chumphon
65. Ranong
66. Surat Thani
67. Phangnga
68. Krabi
69. Phuket
70. Nakhon Si Thammarat
71. Trang
72. Phatthalung
73. Satun
74. Songkhla
75. Pattani
76. Yala
77. Narathiwat

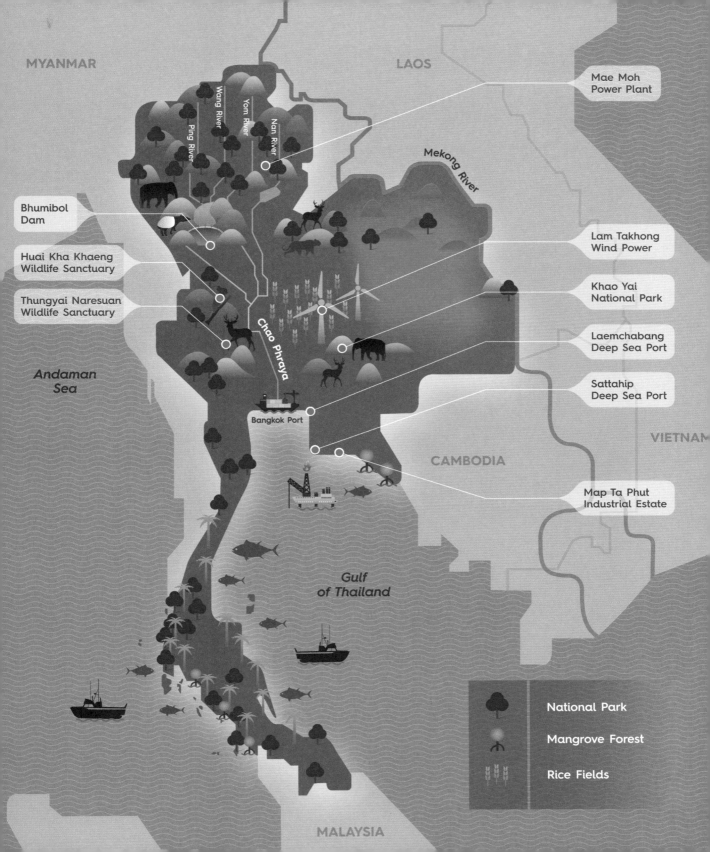

THAILAND OVERVIEW

PEOPLE

Population	68,200,824	Median Age	37.2
Urban	50.4% (% of population)	Life Expectancy	74.7

Source: CIA World Factbook (as of September 2016)

TOTAL LAND AREA

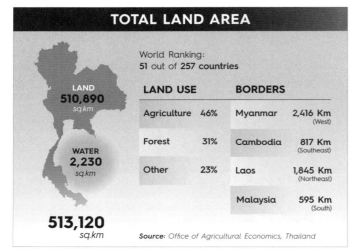

LAND: 510,890 sq.km
WATER: 2,230 sq.km
Total: 513,120 sq.km

World Ranking: 51 out of 257 countries

LAND USE		BORDERS	
Agriculture	46%	Myanmar	2,416 Km (West)
Forest	31%	Cambodia	817 Km (Southeast)
Other	23%	Laos	1,845 Km (Northeast)
		Malaysia	595 Km (South)

Source: Office of Agricultural Economics, Thailand

THAILAND'S ECONOMIC STRUCTURE BY SECTOR

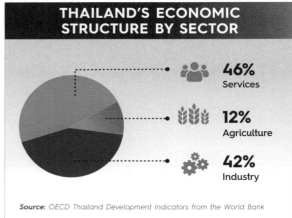

- 46% Services
- 12% Agriculture
- 42% Industry

Source: OECD Thailand Development Indicators from the World Bank

HUMAN DEVELOPMENT BY COUNTRY

(Out of 188 countries)

Thailand is listed among those countries with high levels of human development. Other Asian nations in this category are:

THAILAND: 93

Malaysia	62				
Sri Lanka	73		Philippines	115	
China	90		Vietnam	166	

Source: United Nations Development Programme Human Development Report 2015

ECONOMY

GROSS DOMESTIC PRODUCT (GDP, US$)	$395.28 BILLION
GROSS NATIONAL INCOME (US$ PER CAPITA)	$5,620
INCOME LEVEL	UPPER-MIDDLE INCOME

Source: World Bank 2015

POVERTY

0.6% Population Living Below US$1.90 a day

This is the international poverty line as determined by the World Bank.

0.1% Population in Multidimensional Poverty

Multidimensional poverty identifies multiple deprivations in the same household in terms of education, health and standard of living.

Source: United Nations Development Programme Human Development Report 2015

ASIA-PACIFIC TOP 10 COUNTRIES IN GLOBAL COMPETITIVENESS

Global rankings out of 138 countries

Singapore	2	Malaysia	25	Indonesia	41		
Japan	8	Korea	26	Philippines	57		
Hong Kong SAR	9	China	28				
Taiwan	14	**Thailand**	**34**				

Source: The Global Competitiveness Report 2016-2017

CORRUPTION

76th Out of 168 countries in the 2015 corruption perception index

Source: Corruption Perceptions Index 2015

EDUCATION

7.3 Mean Years of Schooling

in comparison to

13.5 Expected Years of Schooling

Source: United Nations Development Programme Human Development Report 2015

ENERGY

Renewable Energy Production
(MTOE: Million Tons of Oil Equivalent)

26.2 — **12th** rank

Out of 141 Countries
1 = Country with the highest production levels of renewable energy

Share of Renewables in Total Energy Production **33%** — **71st** rank

Source: International Energy Agency Energy Atlas 2016

HAPPINESS

East Asian and Southeast Asian countries in the Top 50 rankings of happiest nations on Earth from 2013 to 2015 out of 157 countries.

THAILAND		**33rd**
Singapore		22nd
Taiwan		35th
Malaysia		47th

Source: World Happiness Report 2016 / Volume 1

THAILAND IS AN AGING SOCIETY

Thailand's population of people aged 65 or older will be equal in number to those aged 15 or younger by **2025**

Source: Thailand Development Research Institute (quarterly review Vol.30 No.1 March 2015)

ACCESS TO THE INTERNET

Internet users (% of population)

34.9%

Source: SDG Index & Dashboards - A Global Report, July 2016

RANKING IN CO_2 EMISSIONS

(Out of 138 countries)
1 = worst

25th

In comparison to other ASEAN nations 2014 data

Cambodia	125
Brunei	109
Singapore	61
Myanmar	48
Philippines	38
Vietnam	34
Malaysia	29
Indonesia	16
Laos	N/A

Source: CO_2 time series 1990-2014 per region/country by EDGAR (Emissions Database for Global Atmospheric Research)

ENVIRONMENTAL PERFORMANCE RANKING 2016

(Out of 180 countries)

 91 Thailand

 1 Denmark

Thailand has improved its track record in areas like Health Impacts (85) and Water Resources (66), but still lags behind many nations in terms of Air Quality (167) and Agriculture (106).

Source: Environmental Performance Index, 2016

Singapore	14
Malaysia	63
Philippines	66
Brunei	98
Indonesia	107
Vietnam	131
Timor-Leste	138
Cambodia	146
Laos	148
Myanmar	153

INTRODUCTION

We live in a digital age where convenience and conspicuous consumption have become a way of life. We are constantly bombarded with advertising for new computers and mobile phones, cars and clothes. New shopping complexes are built all the time, and this culture of consumerism encourages us to spend and live beyond our means. At the same time, our own desires, mindset and behavior, whether manifested through frequent travel or feckless habits, also feed into this cycle of extravagance and waste, bringing us to a precarious place where the planet's population of 7.4 billion people is now using up the equivalent of 1.5 more times the resources than the Earth can replenish in a year.

To check this alarming trajectory, the concept of "sustainable development" has emerged. Around the world, one hears the phrase with increasing frequency. Sustainable development is the subject of international conferences hosted by corporations and governments alike. "Sustainability" is also a contemporary catch phrase for groups across many sectors. The term comes up in everything from tourism to manufacturing to architecture, from governmental economic plans to banking seminars to product marketing. It serves as a modus operandi for NGOs and social enterprises. It appears on pamphlets, business cards and restaurant menus. It is taught in classrooms.

The growing popularity of sustainable development, among individuals, businesses and governments, suggests a paradigm shift may finally be taking hold. As described by Earth Institute Director Jeffrey D. Sachs, sustainable development "is a central concept for our age. It is both a way of understanding the world and a method for solving global problems."

These problems are increasingly complex and varied. Alarm about inequality, food security, climate change, health pandemics, economic crises and conflict is increasing apace with global temperatures – both 2014 and 2015 were the hottest on record – fueling more fervent calls for sustainable development. Big businesses, including those in Thailand, are becoming concerned. As they recognize that they must have dependable human and natural capital to succeed in the future, they are also beginning to acknowledge that they must contribute to the solutions. The fallacy that business is separate from both society and the environment is exposed every time a natural disaster disrupts supply chains, wiping out profits.

Governments, including that of Thailand, have also taken notice. They see sustainable development as a way to organize long-term, multifaceted national development agendas for ever-expanding populations, and as a way to address some of the perennial problems impeding national progress.

It is true that, at times, sustainable development can seem like a catch-all phrase, or, at worst, an empty marketing slogan by those businesses or even governments who use it. But at its heart, sustainable development is a sincere challenge – to create a new mindset that forces us to reconsider how we are living our lives, running our businesses or developing our countries. No organization is dedicating more resources and energy to the cause than the United Nations. The UN sees sustainable development as an encompassing ideology to create more inclusive growth, protect the environment, reduce inequalities, fight climate change, and to reach consensus and action on difficult issues such as greenhouse emissions, human rights and poverty.

In 2015, after many decades of moving the sustainable development agenda into the mainstream, the body convinced world leaders to adopt the strategy as the guiding principle of its new 15-year development program, known as the 2030 Agenda for Sustainable Development. The agenda features 17 Sustainable Development Goals, or SDGs, and 169 targets to help steer countries in the right direction. In the same year, the UN reached the Paris Agreement, which sets out a global action plan to help the world limit global warming well below two degrees Celsius. Representing the first-ever universal, legally binding global climate deal, it was ratified by Thailand in 2016 and will likely come into force by 2020.

Underlying these significant actions is an increasing sense of urgency. Are the leaders of today moving quickly and decisively enough? Is the path we are on still far too unsustainable, set only to lead us toward hardship or even calamity? How can we create the change we imagine? We might first ask, though, how did we arrive at these questions in the first place?

THAILAND'S WAKE-UP CALL

For Thailand the wake-up call undoubtedly arrived in 1997 in the form of an economic implosion. A massive financial crisis saw fiscal institutions go bust, businesses go broke and many people go bankrupt as a bubble economy burst. From its epicenter in the kingdom, the ripples spread across the money markets of Southeast Asia, causing extensive damage.

Over the decade prior to 1997, Thailand had enjoyed spectacular economic growth, frequently in double digits. Inflation was low and foreign investment poured in for infrastructure projects, automobile and electronics manufacturing, textiles and other light industries, as well as property development and the service industries. Tourism arrivals soared. It was a boom time like never seen before in the kingdom's history. All over the country, the signs of this newfound prosperity abounded in the form of shopping malls, golf courses, international schools and hospitals. As Thailand enjoyed some of the best credit ratings in its history, the country looked set to join the ranks of the so-called "Asian tiger" economies.

During this period and the decades leading up to it, social and environmental concerns were rarely raised, which is hardly unusual for developing countries, where providing access to the most rudimentary amenities of life, such as electricity, education and proper sanitation, are the initial priorities. As logging and poaching decimated Thailand's forests and wildlife, tourism led to encroachment on

Bangkok has seen vast development in recent decades with a proliferation of skyscrapers redefining the city's skyline.

national parks, traditional culture was commodified and neglected, and urbanization led to new health and social problems, the question began to be broached: how could the country balance the money-making opportunities of the modern economy with the maintenance of the natural and cultural attractions as well as the social bonds that made it possible in the first place? But who would really listen to such questions when times were so good?

By 1997, however, the baht was overvalued and made an easy target for currency speculators, yields on investments were disappointing, exports were down and foreign debt was sky high. Adding to that volatile mix, the stock market was rife with insider trading and likened to a casino. A panicked withdrawal of credit burst an economic bubble that had been fueled by hot money. Save for a few areas like tourism, much of Thailand's economy was overwhelmed by the crisis. In August of 1997 the International Monetary Fund stepped in with a US$17-billion bailout and mandatory economic remedies that were both unpopular and fiercely debated. As the extent of the damage became apparent – millions were forced out of work by 1998 – there was a national consensus that similar catastrophes had to be averted in the future.

In the midst of this watershed moment, King Bhumibol Adulyadej, who had been the country's head of state since 1946, called for a new mindset, a shift in priorities and a return to a more reasonable, prudent and balanced pursuit of growth. Referring to the extravagance and over-indulgence of the 1990s, he said that "to be a tiger is not important." Instead, he extolled the virtues of sufficiency and moderation, an idea he elaborated on in his birthday speech in 1998 as follows: "Sufficiency is moderation. If one is moderate in one's desires, one will have less craving. If all nations hold this concept without being extreme or insatiable in one's desires, the world will be a happier place."

In a way, this approach and mindset represented the culmination of King Bhumibol's development work, linking many of the major facets together in an overarching framework that puts a distinctly Thai spin on sustainable development. Launched over many decades, these thousands of development initiatives, commonly referred to as the royally initiated projects, were established to address many of the country's most pressing social and environmental concerns: poverty, education, conserving marine resources and forests, developing agricultural cooperatives and implementing water management systems and other schemes to help farming and fishing communities in rural areas who tend to live, at best, a subsistence lifestyle. They engaged directly with the local communities to help better incorporate their local wisdom and insight. They featured affordable technology and environmentally-friendly solutions, such as the building of check dams to fend off floods and irrigate fields, and strategies that took into account the socioecomic conditions and culture of the communities that would be impacted. Long a part of the king's projects, these features have also become hallmarks of sustainable development efforts all over the world.

The king's strategy was ultimately formalized in the late 1990s into the concept of the Sufficiency Economy Philosophy, or SEP. In Thai the philosophy is known as *Setthakit Pho Phiang*. *Setthakit* is the Thai word for economic activities; *pho* is the word for "enough"; and *phiang* means "just". So the phrase means a "just enough economy." After a period of reckless speculation and easy credit that led millions into unemployment, the king's idea, which emphasizes balanced development, captured the national mood at the time.

Although it is often characterized as such, SEP is not a philosophy of austerity, but one that encourages reasonable consumption and expectations. Similar to the sustainable development ethos, SEP does not urge for a return to the past, for Thais to give up all their creature comforts or for development that denies growth or free market mechanisms. It recommends simply that people and likewise businesses live and act within their means. In an article by Professor Harald Bergsteiner and Dr Priyanut Dharmapiya appearing in the book *Sufficiency Thinking: Thailand's Gift to an Unsustainable World*, the framework of SEP is elaborated as "a state of being that enables individuals, families, organizations and nations to enjoy, at a minimum, a comfortable existence and, if conditions permit, a reasonable degree of luxury that balances economic, social, environmental and cultural conditions."

These days, developing such a "moderation mindset" is of paramount importance, given the world's increasingly stretched global resources. Representing one of the three key decision-making principles of SEP (i.e., moderation, reasonableness and prudence), the kind of moderation that the king called for is also synonymous with the "middle way" of Buddhism. It is applied to avoid extremes by trying to balance necessity and luxury, self-deprivation and over-

A Definition of Sustainable Development

What does sustainable development really mean and how does one act sustainably? In the international arena, the most enduring definition comes from "Our Common Future", or the Brundtland Report, published in 1987 by the United Nations World Commission on Environment and Development. It arrived at a time when increasing awareness about the limits of natural resources and the impacts of human actions on the environment had begun to inspire calls for a new, more integrated and balanced model of development. The report stated, quite neatly, that: "Sustainable development is development that meets the needs of the present without compromising the ability of future generations to meet their own needs."

Named after the chairperson of the Brundtland Commission, Gro Harlem Brundtland of Norway, the Brundtland Report focused on redefining the relationship between the environment and development: "...the 'environment' is where we live; and 'development' is what we all do in attempting to improve our lot within that abode. The two are inseparable."

The report was also concerned with securing global equity for future generations by redistributing resources toward poorer nations to encourage their economic growth in a spirit of solidarity that would bring different peoples and cultures together. Among its major insights, "Our Common Future" noted that it's possible to achieve social equity, economic growth and environmental health at the same time. By doing so, it highlighted the three fundamental stakeholders in need of balanced development – the environment, the economy and society. These three components became known as the "three pillars of SD" or the "triple bottom line" in the corporate sense, and they continue to be the key considerations when the concept of sustainable development is applied today.

A CLOSER LOOK AT SEP

The Sufficiency Economy Philosophy (SEP) is a **decision-making framework based on both knowledge and virtues**. SEP is based on the principles of **moderation, reasonableness and prudence**. They translate into appropriate ways to solve problems or take action in different situations.

SEP stresses balance in the use of **economic, social, environmental and cultural capital**, while underlining the importance of preparedness in dealing with changes in these four dimensions. According to SEP, **progress with balance promotes stability and, ultimately, provides a basis for sustainability.** That can be as true for national development programs as for our own agenda in life.

How does SEP translate to "sustainability"?

Decisions are made according to knowledge and virtue:

KNOWLEDGE	VIRTUES
• Insight • Right understanding • Prudent application	• Honesty • Perseverance • Altruism • Mindfulness

Decisions are based on these three principles:

MODERATION	REASONABLENESS	PRUDENCE
• Just enough • Within capacities • Avoiding extremes like overindulgence and deprivation	• Assessing the causes and effects of one's actions on any and all stakeholders including the environment	• Risk management • Preparing for future impacts or change

The outcomes of decisions based on SEP should reflect balanced progress toward sustainability in the four dimensions of life:

Progress with Balance in Life's Four Dimensions

Economic   Social   Environmental   Cultural

Source: Thailand Sustainable Development Foundation

indulgence, tradition and modernization, as well as self-reliance and dependency. Though there are no hard and fast rules about this, it is usually a question of balancing what we have with what we want. The application of Buddhist principles, and the lessons of karma, which show how positive actions beget positive consequences, also provide a Thailand-specific context widely understood by Thai people from all walks of life.

After moderation, the second pillar of SEP is "reasonableness". By this standard we must gauge the impact that our actions and decisions have both on others and the world around us. In the context of sustainable development, it's easy to see how our everyday decisions — such as opting to put garbage into a proper container rather than littering – do or don't translate to reasonableness, because they either solve or create problems.

Another intrinsic part of the philosophy is prudence. As it's defined in the book *King Bhumibol Adulyadej: A Life's Work*, prudence is all about "working carefully, proceeding by stages, growing from an internal dynamic, achieving a level of competence and self-reliance before proceeding further, and taking care not to overreach one's capabilities." Therefore, the decision-making framework of SEP can equally serve the rural farmer or a large business. Finally, decisions based on the principles of SEP must be based on virtue and knowledge. Only then can progress with balance, according to SEP's four dimensions – economic, environmental, social and cultural – be truly achieved.

That the most remembered advice at the time of the Asian Financial Crisis came down from on high was befitting a country that had traditionally looked to its monarchs for leadership and guidance through troubled times – and is suggestive of the uniquely Thai features of its development path as well as the special emphasis that SEP places on the cultural dimension, which is left out of other sustainable development frameworks. Today, SEP has become Thailand's guiding framework for sustainable development.

GROWING PAINS

Before the transition to a constitutional monarchy in 1932, commercial enterprises were nominally under sufferance of the monarch, whose officers regulated everyday life. The tradition of strong centralized control has carried over to modern times with the result that the bonds between government and business remain strong. All the earliest Thai-owned commercial initiatives were started or backed by the government or royalty. Many of

> "Until the manufacturing boom of the 1980s as much as 70 percent of the workforce dedicated themselves to farming."

the descendants of those families and companies continue in business today. Meanwhile, the vast majority of the population worked the land.

Hence, the Thai world view — a world view that encompasses business as well as everything else — revolves around the powerful looking after the weak. Those entrusted with power are expected to wield it with discretion, under obligation to the greater good. Mostly the vulnerable in Thailand have traditionally come from the kingdom's agricultural communities that were — and in some areas still are — the pulse of the Thai heartland.

Until the manufacturing boom of the 1980s as much as 70 percent of the workforce dedicated themselves to farming, a sector which typically accounted for more than 30 percent of the nation's GDP. Farmers' lives revolved around the cycles of the seasons, and the rising and setting of the sun. They used natural fertilizers like manure. They grew their own food. They made their own clothes, tools, baskets and fishing nets. And they also built their own houses, raised up on stilts to prevent them from flooding and to provide a shelter for animals, out of natural materials that were sourced locally. From the forests they picked their own medicinal herbs to use as folk remedies based on local wisdom. Whatever surplus remained from their crops after feeding themselves and sharing with others, they sold or traded for other foodstuffs and supplies. In these ways, many farming families, mostly small leaseholders, were practicing some of the core tenets of SEP and sustainable development long before these terms were coined and popularized.

On top of their agrarian origins, Thais were Buddhists (today, approximately 95 percent of Thais still identify themselves as such), a religion whose teachings and concept of deep ecology overlap in many ways with the principles of sustainable development. The religion holds nature sacred. It stresses moderation, immaterialism and the interdependence of all life forms. Buddhist teachings also reveal the importance of cause and effect, and examine the cycles of life, death and rebirth found throughout the world.

In essence, Thailand's roots and grassroots, its resourcefulness and royally initiated projects as well as King Bhumibol's guidance over several decades could have provided the proper grounding for sustainable development to flourish in the kingdom, but the reality is that it hasn't, which begs the tricky question: why?

As this book discusses, there is no simple answer, but there are some common threads. As former Thai prime minister Anand Panyarachun said in a speech delivered to the Foreign Correspondents Club of Thailand (FCCT) in March 2016, "With the benefit of hindsight, we can see that globalization, consumerism, extravagance, dishonesty and immoderation have led to

> "At this juncture, it has become patently clear that many of our institutions are inadequate when faced with the challenges of globalization."
> – Former prime minister Anand Panyarachun

management failures in both government and business."

"While substantial progress has been achieved in terms of economic development, we have not taken sufficient note of its negative political and social impact. At this juncture, it has become patently clear that many of our institutions are inadequate when faced with the challenges of globalization. Against a backdrop of rapid global change, our economic, political, and social institutions have simply not kept up," he said.

In Anand's view, in addition to heeding the principles of SEP, Thailand requires four key improvements: 1) *sustainable* and *widespread* economic development 2) a more open and inclusive society 3) true respect for the rule of law 4) and a recalibrated balance of power between the state and people. These reforms, he said, can engender the type of responsive leadership and sense of collective empowerment essential to sustainable development.

Citing more specific examples, many observers, including Anand, point to the competing social and business objectives of Thailand's state-owned enterprises as dampening the country's competitiveness and impeding its progress in terms of sustainable development. Thailand's 59 state-owned enterprises, which cover everything from aviation to communications to energy to water to transport, contribute over 40 percent of GDP and have nearly 300,000 employees, giving them a hugely powerful stake in deciding the country's future in key areas and industries. Vulnerable to political interference and corruption, their plans and actions, at times, have left the public crying foul.

Many experts also point to Thailand's educational system, which has not widely produced the requisite critical thinking skills and creativity nor the skilled labor for Thailand to feel confident about being competitive in the 21st century. The centralization of wealth and power in Bangkok is also seen as a perennial problem. While much of the general population finds itself inadequately educated and working in the informal job sector, those with all the advantages of access to wealth and opportunity have watched the gap widen into an abyss as their power grows. A lack of accountability and transparency combined with a culture of impunity from the rule of law by these "haves" has long been a symptom of Thailand's hierarchical patronage structure, one that has left the "have nots" often excluded from enjoying the rights, liberties and opportunities necessary for sustainable development.

As Anand suggests, "Every group, every religion, every region, every rung of society must enjoy these to be able to participate collectively in directing national development. This will instill a critical sense of ownership in the nation's destiny that encourages each and every member of society to keep the state under constant scrutiny."

On the contrary, rampant self-interest, weak institutional integrity and a lack of good governance were all blamed as causes of the 1997 economic meltdown. Until then, one could say that the profit motive had trumped the other democratic pillars of sustainable development in Thailand. But the downturn had an upshot in that it brought these issues into the spotlight.

A man waves a Thai national flag at Democracy Monument in Bangkok, which has been the site of frequent political protests in the last decade.

GREEN LIGHTS

Since the turn of the millennium, however, the sustainable development movement and SEP have begun to dovetail and gain more traction. In Thailand, SEP and sustainable development has begun to inform policy-making, underlined the country's five-year economic development plans and inspired educational reforms in school curricula. In 2007, the United Nations Development Programme (UNDP) devoted its Thailand Human Development Report to explaining the evolution and application of SEP in the public and private spheres. Over the past decade, businesses have started to adopt SEP to guide their corporate governance, human resource development, production processes and – it must be said – boost their PR campaigns. Books, pamphlets, articles and websites have been created to teach the principles of SEP and to publicize the successes of those that follow its principles. Academic studies have been undertaken to accumulate data and help spread SEP into new realms. Through joint cooperation with other developing countries, SEP has even been applied to development projects abroad.

Within Thailand, King Bhumibol's own development work is arguably best encapsulated in the six Royal Development Study Centers, which were established in six key, different agro-ecological areas in Thailand. These centers continue to support research and the creation of development solutions that address the specific challenges facing communities in those regions. Each acts as a "living museum", where farmers and locals can gain and provide knowledge. Over the years, these centers have conducted work on topics such as soil rejuvenation, integrated farming, forest and water management, biofuel development and marine conservation (see graphic on p. 22).

There have also been some examples of strong leadership in the public sector. For decades a key driver of Thailand's manufacturing boom was the building of roads at the expense of railways, in spite of the fact that transporting goods by the former costs more than double the latter. Beginning with the 10th National Economic and Social Development Plan that ran from 2007 to 2011, the road-heavy development model was scaled back in favor of a new emphasis on mass-transit networks around the capital, to be completed in 2022.

One of the main objectives of the current 11th National Economic and Social Development Plan (2012–2016) is to move toward a "low carbon society." But at the same time, state-owned enterprises are still pushing and planning to

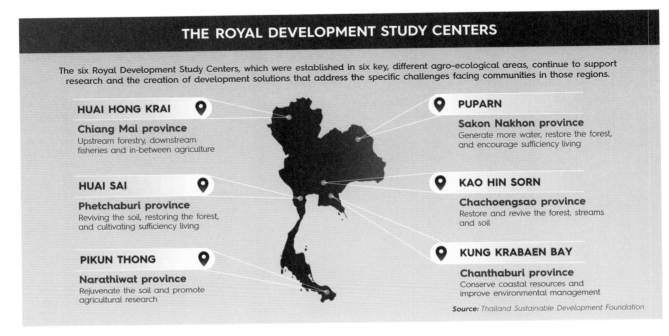

construct coal-fired power plants both at home and abroad. An effort by the Electricity Generating Authority of Thailand (EGAT) to build a coal-fired power plant in Krabi province has come under fire from environmentalists and tourism operators who believe its construction and operations will mar local scenic areas and harm tourism.

In the energy sector, Thailand's current scheme, the Energy Efficiency Development Plan (2015–2036), which is based on tapping a mixed array of sources, calls for more than doubling the country's capacity by 2036 to meet the country's economic needs. But in doing so the government hopes to reduce the use of natural gas to a third of the total while boosting alternative sources by 25 percent over two decades through the Alternative Energy Development Plan (2015–2036). Much of that energy will come from renewable sources like solar, wind and hydro, while the government explores other lesser-known avenues of clean energy such as geothermal and tidal energy.

Reaching that 25 percent goal is not just wishful thinking. Thailand is now the solar powerhouse in the ASEAN region, harnessing more energy from the sun than all the other members combined. The kingdom was also one of the first Asian nations to implement a feed-in tariff, or "adder" program, which offers renewable energy producers long-term contracts to sell electricity at attractive rates. Emphasizing the profit motive in tandem with sustainability, the state is serving as a genuine facilitator for such enterprises, which has been lauded by the business community.

More recently, the government has advanced "Thailand 4.0", an economic model based on creativity, innovation and high-level services. This plan hopes to see the reform of its existing key sectors, such as automotive, electronics,

> "Sustainability has become a benchmark for many corporations that have seen its multifaceted value."

medical and wellness, agriculture, biotechnology and food, as well as the scaling up of new sectors such as logistics, robotics, aviation, and biofuels and biochemicals. The King Bhumibol Adulyadej Institute of Ethics and Education for Sustainable Development will help create progress through vocational and academic training focused on these very sectors, with support from the private sector.

Indeed, on the surface, the Thai private sector is increasingly on board. Sustainability has become a benchmark for many corporations that have seen its multifaceted value. The kingdom's largest oil and gas refiner, Thai Oil, topped the 2016 Dow Jones Sustainability Index List (DJSI) as the Energy Industry Group Leader. All in all, 15 Thailand-based companies were listed on the DJSI as of 2016. For a DJSI-listed conglomerate like SCG, which has built a US$200 million headquarters in Bangkok that is a towering example of "green architecture", it has inspired innovations like a line of 80 different "SCG eco value" products, ranging from cement to chemicals and paper.

When it comes to contemporary culture, the popularity of health-conscious lifestyles has inspired a new wave of small-scale entrepreneurs running restaurants, farmers' markets, organic shops and selling a wide range of artisanal products. Among these are smaller businesses like Sa Paper Preservation House, a family-run business in northern Thailand that applies the principles of SEP and now exports 80 percent of its paper products overseas. Meanwhile, the milk producer Dairy Home has a fair trade agreement with local dairy farmers to make organic products and sell them throughout the country. At a time when agribusinesses are coming to dominate the agricultural sector these are small but significant moves.

In addition, ground-level work in crucial areas of sustainability consists of community-led projects in places like rural Trat province. There, villagers saw firsthand how their mangrove forests had been decimated by shrimp aquaculture, so they set about implementing a new system of checks and balances, with fines for those who disobeyed them, to right this ecological wrong. Another project that started from humble beginnings is the Tree Bank that began in Chumphon province in 2006. Variations on the scheme now exist in all of Thailand's 77 provinces and some 300,000 farmers take part.

The success of such ventures highlights a key driver in every area of sustainable development: the profit motive. Doing good deeds for the sake of the environment is one thing. Doing good deeds while making money is a much more enticing, and ultimately enduring, proposition. What all of these eco-savvy projects have in common, whether they are about sourcing sustainable seafood, transplanting coral reefs or organizing seed banks for farmers, is that not only do they recognize the limits of growth, or what ecologists now refer to as "planetary boundaries", but they also realize that nature is capable of regrowth only if human activities do not adversely affect this capacity, and ecosystems like forests and marine habitats are managed with these considerations in mind.

As such abundant examples indicate, there is a growing awareness of sustainable development in Thailand and more examples of SEP being used as the vehicle to achieve it. Yet clearly there are many challenges. In recent years, for example, big businesses, such as seafood companies, state-owned enterprises and agribusinesses, have been on the receiving end of scathing critiques for supporting a multitude of practices seen as purely profit driven. Examples include the deplorable treatment of fishermen along the supply chain of seafood companies, insider trading by top executives at one of Thailand's largest conglomerates, and the continued facilitation of monocropping at the expense of the forests. The burning of the residue left over from crops like corn, which are grown for animal feed, has led to major incidents of haze and pollution blanketing the north of the country.

Such practices must end if the kingdom is going to become a world-class player in this field. In order to bring about a paradigm shift in consumption patterns and more

Seedlings grown at SCG's Thung Song cement plant.

Hill tribe villagers pick strawberries at Doi Ang Khang.

responsible spending, as household debt in Thailand has risen to 82 percent of the country's GDP, the mindset of the public and consumers must also shift. Many of these challenges are not unique to Thailand and Thailand clearly recognizes some of these key challenges. The recent "Ya Hai Krai Wa Thai Campaign," for example, which can be translated as "Don't let anyone blame Thais," is using communications and public relations to encourage SEP principles and target four key bad behaviors of Thais: littering, corruption and bribe-taking, careless driving and overspending. Indeed, pollution, corruption, road safety and debt were also noted as challenges for Thailand, according to the SDG Index.

How Thailand perseveres and whether or not it makes a strong and sincere commitment to the 2030 Agenda for Sustainable Development is sure to define the country's progress in the decades to come. Just as the 17 SDGs and their 169 targets frame this book's discussion of Thailand's progress, they can provide a road map for the next 15 years and a method to measure progress.

Given its unique framework of SEP, Thailand may have a head start, and it certainly has a compelling vehicle with which it can make progress. In light of the country's achievements in some areas – such as leading ASEAN in solar energy, the creation of many royally initiated projects, companies winning international plaudits and the implementation of an internationally renowned Universal Coverage Scheme (UCS) that gives free medical care to almost all Thais – there are grounds for cautious optimism. Since SEP is now on the curricula of thousands of schools nationwide, the philosophy's most enduring legacy may be yet to come as it influences the thinking, behavior and consumption patterns of future generations of Thais, as they pursue the path of sustainable development.

The 2030 Agenda for Sustainable Development

Without specific targets and goals sustainable development is wishful thinking out of step with reality. The Sustainable Development Goals, or SDGs, adopted by the UN in 2015 are intended to be signposts to show the links between economic, social and environmental issues to better inform international development policies through 2030. They are the follow-up to the Millennium Development Goals, or MDGs, introduced in 2001. Thailand achieved all eight MDGs and is confident of achieving all the upcoming goals too.

The SDGs are more comprehensive and inclusive than the MDGs, even though some critics have questioned where the money will come from to achieve them and suggested that the 17 goals with 169 targets are too unwieldy.

The Sustainable Development Goals are as follows:

1. End poverty in all its forms everywhere.
2. End hunger, achieve food security and improved nutrition and promote sustainable agriculture.
3. Ensure healthy lives and promote well-being for all at all ages.
4. Ensure inclusive and equitable quality education and promote lifelong learning opportunities for all.
5. Achieve gender equality and empower all women and girls.
6. Ensure availability and sustainable management of water and sanitation for all.
7. Ensure access to affordable, reliable, sustainable and modern energy for all.
8. Promote sustained, inclusive and sustainable economic growth, full and productive employment and decent work for all.
9. Build resilient infrastructure, promote inclusive and sustainable industrialization and foster innovation.
10. Reduce inequality within and among countries.
11. Make cities and human settlements inclusive, safe, resilient and sustainable.
12. Ensure sustainable consumption and production patterns.
13. Take urgent action to combat climate change and its impacts.
14. Conserve and sustainably use the oceans, seas and marine resources for sustainable development.
15. Protect, restore and promote sustainable use of terrestrial ecosystems, sustainably manage forests, combat desertification, and halt and reverse land degradation and halt biodiversity loss.
16. Promote peaceful and inclusive societies for sustainable development, provide access to justice for all and build effective, accountable and inclusive institutions at all levels.
17. Strengthen the means of implementation and revitalize the global partnership for sustainable development.

The 169 targets expand on the goals and define them in clearer terms. For instance, under the first goal of ending poverty there are seven targets, including 1.4: "By 2030 ensure that all men and women, particularly the poor and the vulnerable, have equal rights to economic resources, as well as access to basic services, ownership, and control over land and other forms of property, inheritance, natural resources, appropriate new technology, and financial services including microfinance." Besides these targets, the SDGs come with indicators that focus on measurable outcomes. Determining whether or not these goals have been met is up to each government in "setting its own national targets guided by the global level of ambition but taking into account national circumstances", the UN said.

THAILAND AND THE 17 SUSTAINABLE DEVELOPMENT GOALS IN PERSPECTIVE

The following pages take a closer look at the 17 Sustainable Development Goals as well as the opportunities enjoyed and challenges faced by Thailand. While the 17 Goals are treated independently here, in reality they are interrelated. Achieving one will help accomplish the others. Failure to act on one, Goal 13: Climate Action, for example, could prevent the others from being attained. What all the Goals require is a mindset that treats them sincerely and takes them seriously. In terms of the Sufficiency Economy Philosophy (SEP), some Goals emphasize the need for compassion, while others demand a genuine search for innovative solutions. Certainly, none can be achieved without the application of virtue and knowledge. In these pages are some "Calls to Action" and examples of projects that should help show the way toward more balanced and sustainable development that truly protects Thailand's environment, encourages stable economic growth, maintains its social fabric and safeguards its cultural integrity.

THAILAND'S SDG INDEX PERFORMANCE

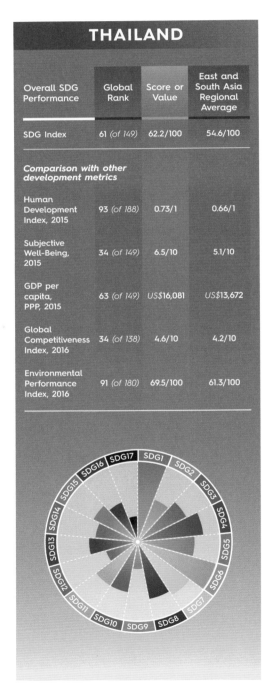

In June 2016, the Sustainable Development Solutions Network (SDSN) headed by expert Jeffery D. Sachs and the Bertelsmann Foundation launched the SDG Index and Dashboards – an annual global report card to track progress on the Sustainable Development Goals (SDGs).

This first SDG Index ranks countries regarding their initial status as of 2015 for each of the SDGs. The index is constructed based on a set of indicators for each of the 17 SDGs using the most recent published data. Indicators have been included that offer data for at least 80 percent of all countries with a population greater than one million. Where possible, the SDG Index uses the official indicators. In cases where official indicators have insufficient data available or where indicator gaps remain, the report authors have reviewed official and other metrics published by reputable sources for inclusion in the SDG Index. For example, Goal 12 – Responsible Consumption and Production – takes into account the percentage of wastewater treated and the number of kilograms of solid waste generated per person per year.

Using the gathered data, an adjusted indicator score that lies between 0 and 100 was generated for each country. This adjusted indicator score marks the placement of the country between the worst (0) and best cases (100). Thailand's score of 62.2, for example, signifies that the kingdom is 62.2 percent of the way from the worst score to the best score.

At the same time, the SDG Dashboards highlight each country's SDG progress across the SDGs and specific targets using a color-coded schema.

Green signifies that for a particular indicator the country is on a good path toward accomplishing an SDG or target, or has, in some cases, already achieved the threshold consistent with achievement. Yellow indicates that the country in question is in a "caution lane", while red means that the country is seriously far from achievement of the SDG as of 2015.

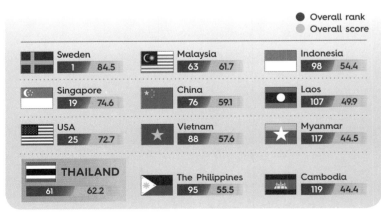

Thailand's Progress on the SDGs

SDG	Thailand's Score	East & South Asia Regional Score
Goal 1 No Poverty	99.91	83.66
Goal 2 Zero Hunger	53.49	47.09
Goal 3 Good Health and Well-being	62.94	61.24
Goal 4 Quality Education	77.60	70.06
Goal 5 Gender Equality	66.01	55.76
Goal 6 Clean Water and Sanitation	94.69	80.97
Goal 7 Affordable and Clean Energy	77.52	58.49
Goal 8 Decent Work and Economic Growth	71.95	59.55
Goal 9 Industry, Innovation and Infrastructure	35.58	24.27
Goal 10 Reduced Inequalities	62.92	68.64
Goal 11 Sustainable Cities and Communities	64.05	56.28
Goal 12 Responsible Consumption and Production	42.73	39.86
Goal 13 Climate Action	58.89	69.84
Goal 14 Life Below Water	50.86	37.29
Goal 15 Life on Land	59.23	46.22
Goal 16 Peace, Justice and Strong Institutions	49.30	56.37
Goal 17 Partnerships for the Goals	29.21	21.18

(Green = achieved / Yellow = "caution lane" / Red = significant improvement needed)

Thailand's Notable Successes and Challenges

	Value	Rating
Poverty headcount ratio at US$1.90 per day (%)	0.1	
Registered births (%)	99.4	
Freshwater withdrawal (%)	13.1	
Access to electricity (%)	100	
Unemployment rate (%)	1.1	
Terrestrial sites, completely protected (%)	53.1	
PM2.5 in urban areas (μm/m^3)	22.4	
CO_2 emissions from energy (tCO_2/capita)	4.5	
Climate change vulnerability (0-1)	0.2	
Traffic deaths (per 100,000)	36.2	
R&D expenditures (% of GDP)	0.3	
Women in national parliaments (%)	6.1	
Homicides (per 100,000)	5	
Prison population (per 100,000)	298	

1 NO POVERTY
\ End poverty in all its forms everywhere.

NO POVERTY
Leaving No One Behind Is Thailand's Next Challenge

Calls to Action

- Decrease household debt by encouraging Thais to save money, spend within their means and take out insurance policies
- Ensure equal access to quality education, skills training and opportunities
- Improve social benefits and safety nets for the poor, disabled and elderly
- Develop proactive solutions to cope with the demographic shift toward an aging society
- Introduce financial safety nets for unskilled workers, small-scale farmers and those who rely on the informal sector
- Scale up programs to improve livelihoods and income generation in rural areas and in female-headed households

The vast majority of Thailand's poor reside in rural areas, and those over age 60 are said to be at particular risk.

Poverty is about far more than how much money a person earns or what they own. It is an obstacle to enjoying opportunities and a decent quality of life. Conversely, by rising out of poverty a person is more likely to have greater access to education, healthcare and other basic services. It also allows them to lead healthier, more productive lives. Unfortunately, across the globe today, 836 million people still live in extreme poverty, while many more balance on a delicate precipice, just earning enough to make ends meet.

For Thailand, such extreme poverty is no longer as pressing an issue. Indeed, according to the SDG Index, Goal 1 is the lone SDG that Thailand has already achieved. The immense economic development the kingdom has enjoyed over recent decades has been generally inclusive, providing jobs and improving standards of living. Initiatives such as the Royal Project (see p. 32) and The Doi Tung Development Project (see p. 36) show how Thailand is committed to **Leave No One Behind** and has even tackled poverty in remote areas, targeting particularly vulnerable populations and providing them with the kind of knowledge and skills that allow them to thrive in their unique local environments without disrupting their traditional culture. This philosophy of helping people to become self-reliant would inform Thailand's poverty eradication efforts for decades to come.

Indeed, Thailand is one of the world's great success stories in terms of poverty eradication, with only six out of 1,000 people (or 0.6 percent

> "The economic development the kingdom has enjoyed has been generally inclusive, but vulnerable groups remain."

of the population) living below the poverty line in 2012. As defined by the World Bank, living below the poverty line means subsisting on less than US$1.90 per day (about 67 baht). Meanwhile, Thailand's Office of the National Economic and Social Development Board (NESDB) defines poverty as living off less than 2,572 baht per head per month (approximately US$2.50 per day). By this measurement, as of 2015, an estimated 7.3 million live in poverty or about 11 percent of the population, a figure that is still relatively low compared to other upper-middle income economies around the globe.

Thailand has also drastically lowered its rates of maternal mortality and infant mortality, among other indicators of human development. A universal health coverage policy has increased the affordability of healthcare and more than 98 percent of the population has access to water and sanitation. By materialistic standards, Thailand also appears to be doing well. In urban areas, the signs of consumer culture are as bright as neon: shopping malls are filled to the brim with luxury brands, omnipresent billboards advertise glitzy condos and car showrooms gleam with the latest deluxe models. Newfound disposable income in the pockets of once-poor laborers, which is often sent back to their families in the provinces, has hastened the rise of more materialistic lifestyles in rural areas as well. So by the time of the economic collapse in 1997, some 90 percent of rural households had a TV set and 60 percent owned a motorcycle.

While this rise in the country's consumerist fortunes has been accompanied by a drop in the overall poverty rate, in reality, these aggregate figures conceal some hard truths. In particular, three groups of people are at the most risk of being left behind by this prosperity: the disabled, single women and the elderly.

Leave No One Behind

A key rallying cry and feature of discussions of the Sustainable Development Goals (SDGs). The idea is that "no goal should be met unless it is met for everyone."

POVERTY IN THAILAND IN PERSPECTIVE

12.7% Of the world's population
0.6% Of Thais
Are Living On US$ **1.90** Per Day (67 Baht)

7.3 Million Thais
11% Of Thais (Approximately)
Are Living On US$ **2.50** Per Day (87 Baht)

- **44.8%** of Thailand's destitute live in the Northeast
- **26.5%** are in the North
- **13.6%** reside in the South
- **80%** live in rural areas

Source: The World Bank, 2015; NESDB, 2015

Thailand's Original Sustainable Development Project

King Bhumibol Adulyadej in the northern highlands in the 1960s.

Long before the term "sustainable development" was even coined, Thailand's monarch, King Bhumibol Adulyadej, was innovating a project that promoted many of the same principles.

Known as the Royal Project, it is an example of a solution to a complex and unique set of problems in Thailand. These problems include border disputes, drugs, poverty among the hill-tribe people who live in infertile areas, and deforestation due to the practice of slash-and-burn farming, which was negatively impacting the region's natural resources.

Thanks to King Bhumibol's vision, many of these complex problems have been alleviated. It all started formally in 1969, when the king established and funded the Royal Project to help increase the feasibility of highland agriculture. The king's intention was to help Thai people living in mountain areas through job creation, to eliminate opium production through crop substitution, and to improve water resource restoration, which incidentally also impacted people living in other parts of the country.

In 1992, King Bhumibol set up the Royal Project Foundation as a legal entity with a solid and efficient management and administration. It was led by HSH Prince Bhisatej Rajani. Describing the project, the prince says: "Following His Majesty's advice, the Royal Project was to find cold weather plants to grow on the mountains. Back then, besides opium, no one knew what to plant. So we started a number of research projects, under which we conducted many experiments. Experiments entail manpower and money. As for manpower, there were many scientists specializing in agriculture who were ready to work for him. His Majesty himself made working under the Royal Project less stressful by curtailing hierarchical procedures and eliminating unnecessary red tape. Research findings from the Royal Project can now be used as findings for teachings. We no longer have to rely on textbooks written by people from other countries."

Currently, the Royal Project Foundation has four research centers and 38 Royal Project Development Centres in Chiang Mai, Chiang Rai, Mae Hong Son, Phayao and Lamphun, working to develop appropriate plants and animal breeds for each specific area, to transfer knowledge

> "Since his first visits to highland villages in 1963, King Bhumibol could see that what ailed the Hmong, Karen, Yao, Akha, Lahu and Lisu communities was not drugs but poverty, poor health and a lack of education."

to local farmers, and to help restore natural resources.

The numbers speak for themselves. More than 350 kinds of plants, vegetables and fruits have been developed. More than 30,000 families or 150,000 people, including farmers from 13 hill tribes and urban areas, have joined the Royal Project, covering an area of 24,000 hectares. With a better management system and a more systematic allocation, more than 450 million baht return to these farmers each year. On average each house is making 70,000 baht annually, which is more than ten times that earned from growing and selling opium. In 2011, there were Royal Initiatives Project Centres covering an area of 272,000 hectares. Some 37,561 families, or 172,309 people, were able to sell agricultural products including vegetables, fruits, flowers, tea, coffee, processed goods and handicrafts of more than 1,702 types to markets in and outside the country, earning a total of 629 million baht.

In rural areas especially, these groups are increasingly vulnerable and isolated. For the elderly in particular, the traditional safety net of being supported by relatives or taken in by the local temple no longer proves adequate in the absence of proper government programs.

There are also considerable and important differences in the incidence of poverty across subnational regions and demographic groups. For example, some 80 percent of the country's downtrodden reside in rural areas, a recent NESDB report noted. About 44.8 percent of the country's destitute live in the Northeast, 26.5 percent in the North, and around 13.6 percent in the South. Almost half of these impoverished households are engaged in the agricultural sector. The rest of the poor are part of Thailand's so-called "informal workforce," comprised of part-time employees, self-employed householders, informal small- and medium-sized enterprises (SMEs), retirees and landless laborers.

In rural areas, it has been found that residents can suffer a "cycle of deprivation". This theory suggests that the key ingredients of poverty, such as bad housing and a lack of access to quality education and employment opportunities are transmitted through families over generations, ensuring that future generations remain in poverty.

In particular, concerns abound about the plight of the Thai farmer. Burdened with mounting debt and losing ground to faceless conglomerates, the so-called "backbone of the country" remains vulnerable to economic shocks, drought, exploitation and natural disasters. According to a survey by the Centre for International Trade Study of the University of Thai Chamber of Commerce, Thai farmers earn the least profits from sales of their crops in all of ASEAN, largely due to higher production costs. As of 2011, the average debt carried by farmers was 104,000 baht, equal to about five years of their income. Clawing their way out of this hole often proves difficult, if not impossible.

Lacking organizational clout, farmers have almost no negotiating power over the trade of their own goods. Terms are instead dictated by the international market price, controlling middlemen and conglomerates. These same conglomerates practice contract farming with farmers, who are often landless (only 29 percent own land titles). By renting land to them, companies cut their costs. In turn, they

2015 AVERAGE HOUSEHOLD INCOME

Northeast 21,094 Baht per month
North 18,952 Baht per month
41,022 Baht per month — Bangkok & 3 neighboring industrialized provinces

Mae Hong Son has been Thailand's poorest province for several years.

Source: NESDB, NSO, 2015

7 Out of 10 of Thailand's POOR LIVE IN THE NORTH AND NORTHEAST

This disabled man became more financially secure by learning to raise mushrooms as a cash crop.

have what amounts to free laborers to work the land, and it is the small-scale farmers who become the most vulnerable to risks such as drought or crop-failure.

Despite Thailand's remarkable progress in joining the ranks of the upper-middle income economies in a brief space of time, the income and opportunity gaps between the haves and the have-nots persist. The question of how to bridge the urban-rural gulf through decentralization or other means and divvy up the spoils of Thailand's larger economic success more fairly and equitably remains one of the country's chronic challenges as it pursues a path toward further development. As it moves toward this goal, it's imperative that the nation creates better opportunities for the poor and middle class to climb the socioeconomic ladder.

EMPOWERING AN AGING POPULATION

These days, declining fertility rates aren't just a phenomenon in wealthy nations. Thailand's birthrate has dipped dramatically from seven children per woman in the 1970s to an average of just 1.5 children today, according to the World Bank. It's also the third-most-rapidly aging society in the world. That makes it one of the less wealthy countries facing the daunting challenge of a shrinking labor pool coupled with a greying population.

In Thailand, the official retirement age for government employees and people employed by state enterprises is 60, while some private sector firms also enforce a compulsory retirement age. Due to a range of factors such as a dearth of surplus income or a lack of planning, the majority of Thais do not put aside savings for retirement. Instead, they rely heavily on their children or on local temples, which offer them refuge.

For decades, a hefty portion of the kingdom's rural working-age population has flocked to Bangkok to seek higher-paying work. Though jobs are abundant, younger Thais are still struggling economically. A consequence of this is that the long-held tradition of Thais caring for their elderly parents has declined. Indeed, the share of older persons relying on children as their primary source of income fell from 52 percent in 2007 to 35.7 percent in 2014.

While it is still common practice for children to send a percentage of earnings home to their parents, the amount often falls short of covering basic needs. Today about half of Thailand's elderly population do not have a child living in the same village or municipality, and 16 percent have no living children to offer support. To make ends meet, a 2014 national survey found that about 38 percent of older people still work regularly. Some 90 percent of these individuals work in the informal and self-employed sector.

In rural areas, the strain is particularly acute, especially among elderly farmers. Community leaders are trying to offset these

Temples in Thailand have often acted as an unofficial safety net for the poor and elderly.

shifts by devising new ways to make the elderly more self-sufficient. One such forward thinker, Mechai Viravaidya, has pioneered a project in Buriram province to help elderly individuals cultivate new skills, increase their productivity, find markets for their products and, as a result, earn more income. And the lessons have certainly taken root. In local villages and on small informal cooperatives, enterprising seniors have adopted innovative techniques such as constraining the roots of lime trees in cement containers in order to make it possible for the trees to yield fruit year round. They've also built homemade shelters, where they grow cash crops like mushrooms and orchids.

The government is well aware of its population's aging issue and has made steady progress on developing policies to support older persons. The Old Age Allowance (OAA) was revamped in 2009 to provide a modest, but nonetheless universal pension to Thais age 60 and over. The government has also supported the establishment of senior citizen clubs that act as self-help groups. By 2012, there were over 23,000 registered elderly groups with a total membership of 1.6 million.

While it's a good start, experts say the government needs to increase spending on existing elderly support and pension systems. And time is running out. By 2040, Thailand's aging population is set to increase to 17 million, accounting for 25 percent of the population. That means that one out of every four Thais will be a senior citizen.

NGOs working on aging issues in Thailand recommend that the government scale up programs that facilitate income-generating opportunities for older people, modify the labor protection law to include older workers, and promote the practice of saving for retirement. Introducing a more flexible retirement age, creating incentives to hire individuals over age 60, and offering flexible hours and part-time arrangements for the elderly would also go a long way toward improving their station. Last but not least, the OAA benefit level should be increased and standardized by linking it to the nationally defined poverty line, and it should be regularly adjusted to reflect inflation.

A woman rests after giving birth at her home in Yala province with the help of a traditional midwife.

AN ALARMING ACCELERATION OF HOUSEHOLD DEBT

A growing concern in Thailand is that the kingdom's poor and middle class are saddled with significant household debt. While the economic boom of the 1980s and early 1990s

The Doi Tung Development Project

Doi Tung was once a shady and secluded area in the black heart of the Golden Triangle, which had earned the dubious distinction of being the world's largest opium-producing region. But that was far from the area's only problem. For decades this watershed had been denuded by slash-and-burn cultivation. The downtrodden locals, composed of members of six different ethnic groups, lived in dire poverty with little access to running water, electricity, healthcare and schools. To make matters worse, the area was a hotbed of armed militia and cutthroat opium barons, who made it even more difficult for government officials to provide any assistance to the hill tribes.

That is the backstory of and the catalyst for the Doi Tung Development Project (DTDP), a highly successful crop substitution program and one of the Mae Fah Luang Foundation's four flagship projects. First established in 1988 under the royal patronage of Princess Srinagarindra, the mother of King Bhumibol Adulyadej, the project area covers approximately 15,000 hectares and benefits around 11,000 people from 29 different villages around Doi Tung Mountain in Chiang Rai province.

Understanding the interconnected nature of issues like poverty, health, hunger, employment, security and land management, the "Princess Mother" developed a multi-pronged plan that respected the local environs, took into account the local culture, wisdom and social conditions of the villagers, and aimed to boost their livelihoods. As a new cash crop for the farmers, she chose Arabica coffee trees. Though not indigenous to Thailand, this variety of tree flourishes under the shade of the forest. This meant that the forest would not have to be cleared to grow it.

The long-running endeavor is now the most globally renowned of all the royally initiated projects. The crop substitution program, first pioneered by King Bhumibol in 1969, has inspired similar ventures in opium-plagued countries like Myanmar and Afghanistan, although Thailand remains the world's most successful example of eradicating opium production at its roots.

The timeframe for the DTDP is 30 years, which has been broken down into three phases. During Phase I, which ran from 1988 to 1993, the first priority was tackling health issues and providing vocational training. From 1994 to 2002 the project's linch pin was income generation, as the DTDP introduced the concept of moving up the value chain by building a factory to roast the beans and package them under their own Doi Tung brand. The final phase, set to finish in 2017, is about strengthening the business units so that the brand and the community are sustainable, as well as capacity building and education, so that locals can take over the project when it concludes.

To ensure a smooth transition from the old guard to the new entrepreneurs, the DTDP has collaborated with the Ministry of Education on reforming the curriculum in line with international standards to provide opportunities to study and hands-on vocational training programs that are adapted to a local context. The programs are also designed to instill students with cultural pride and provide them with enough skills to find good jobs locally so they can become productive members of the community.

Now that the Doi Tung brand has expanded to encompass four different business units (i.e., food, handicrafts, horticulture and tourism), it provides expanded career opportunities at their two resorts, including the Greater Mekong Lodge on the grounds of the Hall of Opium, a high-tech museum in the Golden Triangle Park, and at their restaurants and cafes with outlets in Bangkok, Chiang Mai and Chiang Rai. The Doi Tung Lifestyle Shops in those three cities also sell a range of hand-woven clothes, carpets, mulberry paper, ceramics and home décor.

Recognized the world over as a paragon of sustainable alternative livelihood development, all of the project's products sport the seal of the United Nations Office on Drugs and Crime as a hallmark of its success and humble origins. The area is also notable for being a beacon of cultural, social, economic and ecological development that gives equal weight to all of these considerations and blends them into a harmonious whole.

ushered in a new era of prosperity, on the downside a greater access to credit, coupled with a series of demand-side economic stimulus measures by the government, also drove up the level of household debt.

A long period of accommodative monetary policy implementation and intensified competition among financial institutions has also been partly responsible for the rapid rise in household debt, which surged to 84.2 percent of GDP in 2015, up from 54.6 percent of GDP in 2007.

In more recent years, the slower GDP growth and the government's tax incentives for first-home buyers and first-car buyers to spur domestic demand have also played significant roles in pushing up the debt level of Thai households. Some 36.7 percent of that debt was owing to the purchase of land and residential units while around the same amount was spent on household consumption. The remainder was spent on business, farming, education and other expenditures.

According to a survey on the economic and social conditions of the household sector by the National Statistical Office, average household debt leapt to 156,770 baht in 2015 from 116,681 baht in 2007. This distressing acceleration of household debt threatens future liquidity and the debt serviceability of the household sector, and also raises serious concerns about nonperforming loans at financial institutions. The continuing increase in household debt could also create more adverse effects as it impinges on purchasing power, limits consumption to the bare necessities and drags down the overall standard of living.

Although there have been no imminent signs that the loan quality at commercial banks is deteriorating while household debt grows at a slower pace, the Bank of Thailand has periodically expressed its concerns over these high figures. The situation needs to be closely

Employees interact with customers browsing Apple Watches at an iStudio store in Bangkok.

monitored by financial institutions and regulators who must take steps to ensure that household debt does not go through the roof.

There is no need to be too alarmist; precautions can be taken. In order to strengthen the financial position of Thai households and increase their immunity to future economic risks, it is essential to promote fiscal discipline by reducing excess spending while at the same time encouraging people to save money, stop borrowing and take out insurance policies to cope with emergencies. Shoring up the finances of the household sector would enhance overall economic stability, too. It would keep the wheels of the economy turning by ensuring that the public's purchasing power is not impeded, thereby mitigating the risk of another recession over the years to come.

Further Reading

• *Poverty, Income Inequality, and Microfinance in Thailand* by the Asian Development Bank, 2011

• *Statistical Yearbook for Asia and the Pacific 2014* by the United Nations Economic and Social Commission for Asia and the Pacific

2 ZERO HUNGER

End hunger, achieve food security and improved nutrition and promote sustainable agriculture.

ZERO HUNGER

Ensuring Food Security by Creating a Sustainable "Kitchen of the World"

Farmers at work in the rice fields of Thailand's central plains.

Calls to Action

- Create a mindset among farmers and agribusinesses that emphasizes balance between the profit motive and care for the environment
- Replace monoculture, deforestation, over-grazing and chemically dependent agriculture with sustainable, self-reliant farming practices
- Adapt agriculture to the impacts of climate change, and reduce its contribution to climate change
- Reduce the rate of stunting
- Decrease contract farming; make farmers less reliant on agribusiness and more resilient to risks

Food security, often defined as the state of having enough nutritious food to promote healthy lifestyles, has been a prime concern for communities worldwide since the beginning of time. But today, we're facing new challenges to this age-old problem. While climate change and a global population boom pose new threats to food security, our heavily industrialized methods of food production are now contributing to climate change and environmental harm, rather than adapting to or mitigating it. In fact, food production is the single most important driver of environmental harms around the world.

With a well-earned reputation as the "Kitchen of the World", Thailand has a significant role to play in the field of food security and food production. Counted among the world's top producers of food, Thailand produces roughly a third of the world's stock of rice. It is also the only net food exporter in Asia, as well as one of the world's leading producers of cassava, chicken, eggs, tropical fruit, canned tuna and frozen seafood. Meanwhile, Thailand's agriculture sector employs approximately 40 percent of the country's workforce, providing livelihoods for roughly 25 million people. How Thailand rises to the challenge of developing sustainable food production methods will not only impact millions of Thai lives, but also the environment at large and the nutrient value of the food exported to the rest of the world.

Agriculture – particularly rice cultivation – has been the lifeblood of the Thai people for millennia. Although the country boasts a rich heritage of traditional and sustainable farming practices, the majority of all farms in Thailand became **monoculture** during the Green Revolution of the 1960s and 1970s. This more mechanized, chemical-dependent and unsustainable method of farming increased productivity and made Thailand a major agricultural exporter. However, the Thai landscape suffered severe side effects: soil erosion, deforestation, dwindling biodiversity, and over-reliance on chemical pesticides and fertilizers.

Thai farmers have shouldered the brunt of this agricultural transformation. Bought out by agribusinesses that thrive on monoculture, most Thai farmers today are either small-scale

> "While climate change and a global population boom pose new threats to food security, our heavily industrialized methods of food production are now contributing to climate change and environmental harm, rather than adapting to or mitigating it."

land owners or landless (only 29 percent own land titles) and have taken up **contract farming** with conglomerates. Obligated to produce a certain amount and type of crop each season, but also relying on these same conglomerates to provide everything from seeds and feed to chemicals and equipment,

Food security
In 1966, the World Food Summit defined food security as "when all people, at all times, have physical and economic access to sufficient, safe and nutritious food to meet their dietary needs and food preferences for an active and healthy life."

Monoculture
Also known as "monocropping", monoculture is an agricultural practice that devotes vast tracts of land to the cultivation of just one crop. Monoculture often requires the use of pesticides and chemical fertilizers due to the lack of biodiversity on the farm, which usually keeps pests at bay.

Contract farming
A type of agricultural production in which the buyer and farmer reach an agreement stipulating the price and/or quality and time frame for a specific amount of an agricultural product to be delivered. Normally the buyer is an agribusiness that provides such things as seeds, fertilizers and technical knowhow on credit to the small-scale farmer.

farmers have become ever more dependent on chemical inputs and fallen deeper into debt. Today, not only do farmers make up the poorest population of Thailand, but they also suffer from health issues caused by exposure to chemicals.

And that's just the beginning. Farms are major contributors of the three most harmful greenhouse gases, CO_2, methane and nitrous oxide. In Thailand specifically, the agriculture sector's over-use of herbicides, pesticides and fertilizers have sapped the fertility out of arable land, contributed to water pollution and depleted biodiversity. Intensive irrigation has depleted underground water sources that

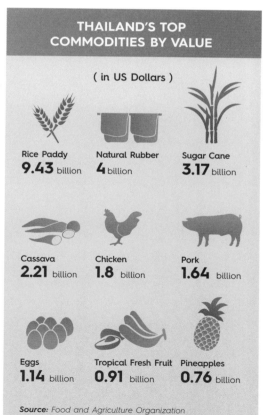

THAILAND'S TOP COMMODITIES BY VALUE

(in US Dollars)

Rice Paddy	Natural Rubber	Sugar Cane
9.43 billion	**4** billion	**3.17** billion
Cassava	Chicken	Pork
2.21 billion	**1.8** billion	**1.64** billion
Eggs	Tropical Fresh Fruit	Pineapples
1.14 billion	**0.91** billion	**0.76** billion

Source: *Food and Agriculture Organization of the United Nations, 2012*

New Theory: A Uniquely Thai Form of Sustainable Agriculture

A farmer who follows King Bhumibol Adulyadej's New Theory method.

Concerned about the security and struggles of Thai farmers, especially in remote and drought-stricken areas, King Bhumibol Adulyadej formulated his "New Theory" farming system in the mid-1990s primarily to alleviate poverty. As such, the theory addressed the pressing issues of land and water management, crop diversification, and promoted self-sufficient farming methods. However, seen in a more contemporary context of food security during an era of climate change, the New Theory framework also offers a uniquely Thai form of sustainable agriculture with a local, cooperative and integrated approach. With its prudent, knowledge-based approach to development and sensitive, sensible allocation and use of resources, the New Theory bears many of the hallmarks of SEP thinking.

The idea came about almost serendipitously when King Bhumibol's Chaipattana Foundation began to develop land in Saraburi province. The initial plan was to build a community center where people could learn new farming techniques. King Bhumibol suggested a pond be dug to store water for the dry season. When the king visited the Wat Mongkol Chaipattana Area Development Project, he approved of the way the land had been divided: 30 percent was allocated to the pond, 30 percent to rice paddy, another 30 percent to mixed crops and 10 percent to other household uses. This formula became the basis of his New Theory.

For practical implementation of the New Theory model, King Bhumibol proposed three stages of development:

STAGE ONE - Family-oriented self-sufficiency: During the initial stage the goal is for the family to have enough to live on, with good shelter and security. Thus the land is divided into four parcels in a ratio of 30:30:30:10. One 30-percent parcel is made into a pond to store rainwater for use during the dry season. Aquatic animals and plants such as fish, shrimp and morning glory may be raised in the pond for food. Another 30-percent parcel is for rice farming for consumption. The other 30 percent is for growing fruit, herbs and vegetables for consumption and, in case of a surplus, for marketing. The remaining 10 percent of the land is for a residence, a corral for livestock, and other buildings.

STAGE TWO - Community-based agriculture: Farmers can then join hands and set up a cooperative to prepare soil, distribute seedlings and make an irrigation system among other activities. Cooperatives can also market products and provide loans and scholarships to members.

STAGE THREE - Reaching out to external markets: Farmers or cooperatives may find financial sources such as banks and private firms to provide funding to distribute or export goods outside the local community.

Many communities have come to their own conclusions that monoculture is no longer an economically viable option for them. As a result, they have decided on their own to engage in a form of diversified farming in line with King Bhumibol's New Theory method. Many such communities now serve as learning centers for others who want to acquire knowledge about integrated farming.

cannot be replenished. Forests have been razed to make way for monoculture plantations, destroying natural habitats as well as valuable carbon sinks. Slash-and-burn practices to make room for corn and plantations have polluted the air. Meanwhile, the planet's growing reliance on a diminishing number of large-scale, monoculture farms threatens food security. Without seed diversity or varietals, extreme weather events caused by climate change can wipe out entire crops, leaving scant alternatives for nourishment.

If you think this sounds bleak, you aren't alone. But you can take heart in the many promising efforts being made in line with SEP that suggest a change in mindset is possible across the Thai agricultural sector.

SHIFTING TOWARD MORE SUSTAINABLE AGRICULTURE

In the mid-1980s, a loose amalgamation of NGOs, small-scale farmers and cooperatives joined to form the Alternative Agriculture Network to promote sustainable agriculture in Thailand. They defined "sustainable agriculture" as a more holistic approach toward "agricultural production and farmer livelihood

INTEGRATED FARMING IN THAILAND BY THE NUMBERS

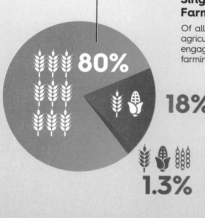

Single-product Farming Monoculture
Of all Thailand's 5.9 million agricultural holdings, 80% engage in single-product farming monoculture.

18% Two & Three Commodities
Of the other fifth, around 18% have two commodities while a mere 1.3% boast three or more different products.

In total, 78,903 farms simultaneously cultivate crops, rear livestock and culture fish in fresh water. The majority of them (56,910) are located in the country's Northeast (38,695) and in the North (18,215).

Nearly half (46.7%) of all argricultural holdings are in **THE NORTHEAST REGION** of Thailand, including the greatest number of multiple-product farms, with about **one quarter offering two products and just over 1% producing three or more commodities.**

33% of all argricultural holdings producing freshwater fish are located in **THE NORTHEAST**, the highest total in the country.

Source: Thailand's 2013 Agricultural Census

Oil palm plantations have been expanding in Thailand at around nine percent per year.

that contributes to the rehabilitation and maintenance of ecological balance and the environment, with just economic returns, promoting a better quality of life for farmers and consumers...for the benefit and the survival of all humankind." Today the Alternative Agriculture Network has helped launch Thailand's nascent organic movement as well as community-based sustainable farms and several fair trade initiatives.

Thailand is also moving toward improved food security and food production by taking more nuanced, region-specific approaches toward farming. According to Jeffrey D. Sachs, director of the Earth Institute, the best way to ensure global food security is to take a local approach, adapting sustainable systems to local environments for precise, region-appropriate food production. In line with SEP, Thai policymakers are already advocating just that by promoting integrated farms as a key objective in the 11th National Economic and Social Development Plan (2012 to 2016). A holistic farm management system that incorporates multiple species of produce and livestock tailored to the needs of local ecosystems, integrated farming ensures sustainability for both the land and the people who work it. But, in fact, this is nothing new in Thailand – integrated farming has been practiced in one form or another for centuries. According to the National Economic and Social Development Board, integrated farming is still the most common method of sustainable agriculture practiced in Thailand.

Although industrialized agriculture and food production has already deeply damaged Thailand's natural landscapes and farming communities, today there are many exciting sustainable methods that are beginning to be implemented. Surely, with a government-supported push toward sustainable agriculture, and with a variety of existing sustainable farms from which to learn, Thailand is in a good position to improve food production – and, as the "Kitchen of the World", to have far-reaching, global impacts.

COMBATTING MALNOURISHMENT AT HOME AND ABROAD

As a top food producer that feeds millions of people across the globe, Thailand has both the opportunity and the obligation to provide the international market with healthy, nutrient-rich foods. One way Thailand is rising to the occasion is through the promotion of organic food production. Organics not only reduce the body's exposure to harmful toxins, but research also suggests that organic foods offer more nutrients such as protein, minerals and vitamins than non-organic foods. And because children are especially susceptible to the dangers of pesticides and other chemicals, many nutritionists and pediatricians argue that organic foods promote better health and development in children.

Emerging over the past few decades as a reaction against agribusiness, Thailand's nascent organic movement is gaining momen-

> "The best way to ensure global food security is to take a local approach, adapting sustainable systems to local environments for precise, region appropriate food production."

THAILAND PRODUCED MORE THAN *2.7 Million Tons of Rice in 2015–2016*

The Maha Yu Sunthornchai Method of Integrated Farming

Maha Yu Sunthornchai was a humble farmer who hailed from Surin province in the northeast, or Isan, the largest and poorest part of the country. Agriculture is the main livelihood in this area, which is afflicted by drought and poor soil.

Through his own observations and patient practice, Maha Yu developed a form of integrated farming specific to this region that restores soil fertility, provides plenty of food and household items for personal usage, alleviates poverty and increases resilience to climate change. His farming model is now widely used in Surin and beyond.

Aerial view of Maha Yu's farm in Surin province.

In 1947, Maha Yu inherited about seven hectares of land. At the time, Thai agriculture was transitioning from subsistence farming to agribusiness by employing new technologies and chemicals. Maha Yu sought a way to make his farm more in line with Thailand's new goals of productivity and profit without stripping the region's environment of its scarce resources or sacrificing his financial independence as a farmer.

Until roughly 1960, Maha Yu practiced a form of mixed agriculture that he had learned from his parents. The practice involved selecting the best seeds and varietals for the farm's conditions in order to ensure good yields despite unpredictable weather, labor shortages and fluctuations in commodity prices. Yet he still cultivated these crops separately.

It was not until 1970 when he visited an integrated farm where fish, rice and pigs were raised together that he discovered that diversity for its own sake does not necessarily yield ideal results; rather, it's how these elements feed and play off each other that drives down costs and increases fertility. While fish swam in the rice paddies and fertilized the soil, rice husks fed the pigs, and the pigs' manure provided the fish with food. Meanwhile, fishponds collected rainwater and maintained the moisture that was badly needed in the arid land. The end result was a self-reliant, high-nutrient, low-cost, sustainable farm that recycled resources and did not rely on expensive chemicals.

Upon returning home, Maha Yu dug fishponds in his rice fields and began experimenting with the twinning of agriculture and aquaculture. According to Maha Yu, a well-balanced integrated farm can provide every need if a one-rai plot incorporates eight essential commodities: rice, fish, pigs, poultry, vegetables, fruits, herbs and medical plants. This formula is beneficial in several ways. First, the farmer's independence from chemicals, machines and other external inputs helps cut costs. Second, the farmer sells only surpluses, whatever he and his family cannot consume, giving the household greater autonomy over what is consumed and what is sold. Third, the byproducts from the agricultural commodities are used to improve productivity, as well as to regenerate nutrients and resources. Fourth, the localized and largely chemically free practices of the farm reduce its environmental impacts. Finally, the greater diversity of farm products increases food security and mitigates risk from natural disasters.

Built on long-standing local and traditional knowledge, Maha Yu's model of integrated farming also embodies some of the tenets of the King Bhumibol Adulyadej's New Theory and Sufficiency Economy Philosophy and his farm now acts as a learning center for other farmers interested in adopting these principles. In 2002 he founded a farmer's group called Local Wisdom of Isan. Judging by the fact that the region now has the highest concentration of integrated farms in the country, Maha Yu has been both an inspiration and an evangelist.

The Global Hunger Index

Believe it or not, roughly 40 percent of the world's population suffers from malnourishment. This comes in three forms: undernourishment (also known as chronic hunger), micronutrient insufficiency and obesity. While the first two are forms of hunger that plague developing nations, the third, obesity, is a form of malnourishment that primarily plagues developed and wealthy nations.

Putting obesity aside, the Global Hunger Index (GHI) combines four factors into one index to measure hunger around the world: undernourishment (the percentage of the population who get an insufficient caloric intake), child wasting (the proportion of children under the age of five who suffer from low weight for their age), child stunting (the proportion of children under the age of five who suffer from low height for their age), and child mortality (the mortality rate of children under the age of five). The GHI assigns hunger scores on a 100-point scale. Zero is the best score, reflecting zero hunger, while 100 is the worst score.

In general, Thailand has seen vast improvement in its hunger index, with the country's remarkable economic growth over the last 25 years making the country a regional leader in development. Thailand has drastically decreased its GHI score from 28.4 in 1990 to 11.9 today. However, the country still ranks 37th out of the GHI's 104 developing countries, revealing that there's room for improvement.

According to Thailand's Multiple Indicator Cluster Survey (MICS) 2012, which was carried out by the National Statistical Office with assistance from UNICEF, one child in six under the age of five is stunted. In addition, nearly one in 10 children under age five is underweight and 6.7 percent suffer from the acute level of undernourishment known as wasting.

ORGANIC FARMING IN THAILAND BY THE NUMBERS

The Predominant Organic Products in Thailand Are

Rice   Vegetables   Fruit

which represent about **0.2%** of Thailand's total farmlands.

Land Use for Organic Farming Between 2000 and 2014

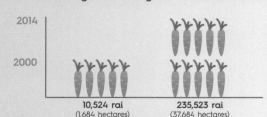

10,524 rai (1,684 hectares)   235,523 rai (37,684 hectares)

Three Channels Where Organic Products Are Sold

Conventional Supermarkets   Specialty Shops   Direct Supply such as farmers' markets or cooperative memberships

Source: Thai Organic Trade Association, Green Net, Earth Net

Estimated value of organic produce in Thailand in 2014
2.3 billion baht

tum. In the 1990s, the Alternative Agriculture Network established a domestic certification body, the Organic Agriculture Certification of Thailand (ACT). From this point on, organic farming has been on the upswing, increasing from roughly 1,000 hectares of land under cultivation in 1998 to 37,684 hectares in 2014.

Most of Thailand's organic crops are produced for export, meaning that Thailand's healthiest foods are feeding an international rather than a domestic market. The main crop is rice, while other cash crops, such as soybeans, peanuts, tropical fruits, asparagus, tea, coffee and herbs bulk out the total. However, the domestic interest in and market for such food products remains relatively low.

> "Less than 0.2 percent of all arable land in Thailand is under organic cultivation today, and only 0.15 percent of all farming households were certified organic in 2011."

Also, with the vast majority of Thai agriculture still being produced under monoculture, organics have a long way to go. Toxic pesticides and other chemicals long abandoned by more developed countries are still used extensively here. According to Green Net, an organic agriculture research agency, less than 0.2 percent of all arable land in Thailand is under organic cultivation today, and only 0.15 percent of all farming households were certified organic in 2011. Thankfully this number is on the rise, but organic advocates maintain that greater demand at home is needed to scale up organic food production in any meaningful way.

According to sustainable development consultant Jeff Rutherford, the organic movement in Thailand is largely dependent on the individual purchases of health-conscious consumers, which makes for small sales and slow progress for the movement as a whole. But a look at Europe, the US and Japan shows a growing institutional demand for organic foods and products. In Thailand, this untapped market of institutions, such as municipal bodies, government departments, schools and corporations, has the potential to scale up demand in a major way by contracting organic farmers to supply entire institutional populations.

With existing certification infrastructures and government support, organic farming has real potential in Thailand. Making more inroads into marketing and distribution, as well as forging more partnerships between farmers, consumers and institutions, can boost domestic interest. This in turn may bolster Thailand's organic movement, providing healthier foods both at home and abroad.

THAILAND RANKS
37th in the Global Hunger Index

Further Reading
- *Overview of Contract Farming in Thailand: Lessons Learned* by Asian Development Bank Institute, 2008
- *Thailand's Progress in Agriculture: Transition and Sustained Productivity Growth* by Henri Leturque and Steve Wiggins, Overseas Development Institute (ODI), 2011

3 GOOD HEALTH AND WELL-BEING — Ensure healthy lives and promote well-being for all at all ages.

GOOD HEALTH AND WELL-BEING

An Abundance of Success Stories Yet Much Room for Improvement

A free aerobics program in Yala. An emphasis on prevention is needed for Thailand to improve its public health.

Calls to Action

- Reduce inequalities between rural and urban areas in the distribution of health resources and medical personnel
- Focus on prevention by encouraging Thais to eat better, exercise more and take care of their spiritual well-being
- Strengthen the prevention and treatment of substance abuse
- Improve road and driving safety to reduce accidents
- Improve treatment and care for people suffering from mental illness
- Promote a more caring, sharing society

Promoting healthy lifestyles and ensuring that each and every person has access to healthcare is essential to sustainable development. Globally speaking, significant strides have been made in increasing life expectancy, curbing the spread of HIV/AIDS, and reducing some of the common killers associated with child and maternal mortality. However, better efforts are needed to fully eradicate a wide range of diseases and address the many persistent and emerging global health issues we face today.

Overall, Thailand has a good head start on achieving Goal 3. In 1930 the average life expectancy in Thailand was a mere 31 years. But by the 1950s it had risen by some two decades, and as of 2015 it stands at 74 years. That leap in longevity speaks volumes for the vast improvement of the kingdom's healthcare system. Thailand had achieved most of its health-related Millennium Development Goals by 2004, well in advance of the 2015 deadline laid down by the United Nations in the year 2000. Among these triumphs is Thailand's positive maternal mortality and neonatal mortality rates; infant vaccination rate (99 percent) and healthy life expectancy at birth; the elimination of malaria in all but the most far-flung frontiers; and the cutting of new HIV infections by more than 80 percent since the peak of the pandemic in 1991. Furthermore, almost 98 percent of the populace now has access to improved water and 93 percent enjoy proper sanitation.

Meanwhile, Thailand's School Lunch Program, the origins of which date back to the 1950s, has been crucial in combatting malnourishment. By giving schools a 20-baht subsidy per beneficiary, the program can provide one meal per day to around 2.5 million malnourished and impoverished children each year in some 30,000 public primary schools across the kingdom.

Such headway is not surprising given Thailand's long history of both traditional medicine based on local wisdom and modern medicine, and its more recent track record of facing down threats and improving its facilities, sanitation, water treatment and professional know-how. As far back as the 1800s, Thailand's kings were supportive of Christian missionary-led health clinics and services, so much so that all missionaries were referred to as "doctor". In the 1900s, the royal family continued to support the expansion of the medical system, with King Bhumibol Adulyadej's own father, Prince Mahidol, leading the way through the modernization of Siriraj Hospital in the 1920s.

In the 20th century, as Thailand's medical system modernized and global breakthroughs in prevention and treatment were adopted, outbreaks of polio, smallpox, cholera, yaws and other deadly diseases, as well as the scourge of malaria, were largely eliminated as rural and urban menaces. However, 21st-century Thais face a slew of new threats including obesity, substance abuse, poor road safety and the health ramifications of living in an increasingly industrialized environment. In addition, according to the chapter "Sufficiency in Thai Healthcare" in the book *Sufficiency Thinking*, "a shortage and maldistribution of health personnel, together with skilled health professionals shifting from public to private hospitals and from rural to urban areas, have contributed to inequalities in health outcomes between rich and poor." Indeed, an estimated 50 percent of all physicians practice in and around Bangkok.

Compounding these issues, the soaring interest in medical tourism, specialty medicine and services like cosmetic surgery is driving both doctors and hospitals to chase profits. A two-tiered system has therefore developed: an expensive, profit-driven one for the rich, and an understaffed, less efficient one for the masses. Neither one of these is focused enough on the preventive, patient-centered and holistic care that would truly alleviate pressures on the healthcare system and make it more sustainable in the long run.

HEALTHCARE COVERAGE

- 8% **Civil Servant Medical Benefit Scheme** — 5 million civil servants, including their spouses and children under 21
- 17% **Social Security Scheme** — 10.77 million private sector employees
- 75% **Universal Coverage Scheme** — 48.61 million people, including children, the elderly and disadvantaged

Source: Public Health Ministry, 2013

LEADING CAUSES OF DEATH IN THAILAND

19%	Cancer
12%	Heart disease
10%	Strokes
9%	Respiratory infections
4%	HIV/AIDS

4%	Chronic pulmonary diseases
4%	Diabetes
4%	Road accidents
2%	Kidney disease
2%	Cirrhosis

Source: Public Health Ministry, 2012

Kingdom Eliminates Mother-to-Child Transmission of HIV

Thailand has had considerable success in preventing the spread of HIV, and has been called a model for other countries to follow after promoting 100 percent condom use among sex workers in the 1990s. In June 2016, Thailand achieved another key milestone when the WHO announced that it had become the first country in Asia to effectively eliminate mother-to-child transmission of HIV and syphilis.

The number of babies contracting HIV from their mothers in the kingdom fell from over 3,000 in the late 1990s to less than 90 in 2015. While that figure does not represent 100 percent eradication, it meets the WHO's criteria by reducing the mother-to-child transmission rate to less than two percent and fewer than 50 new infections in 100,000 births.

Credit for Thailand's success can be attributed to vigilant screening for HIV during prenatal care, and treatment with antiretroviral drugs for those women who test positive. This is significant because if left untreated, HIV-positive mothers have a 15 to 45 percent chance of transmitting the virus to their babies. But that figure plummets to about one percent when antiretroviral drugs are given to both mother and child throughout the stages when infection can occur. According to the Ministry of Public Health, 98 percent of pregnant women living with HIV have access to antiretroviral drugs and almost 100 percent of babies born to HIV-positive mothers are given antiretroviral treatment. Across the nation, concerted efforts to raise awareness about the dangers of HIV/AIDS have also helped reduce infection among women of childbearing age.

It is worth bearing in mind, though, that hundreds of thousands of female migrant workers from neighboring countries like Cambodia and Myanmar are not included in the data. These women have limited healthcare access in Thailand, and many are reluctant to get tested or treated for HIV due to concerns that they will lose their jobs or risk run-ins with the police or immigration authorities. However, a 2010 government study found that in some areas as many as two to three times more migrant women were infected with HIV than their Thai counterparts. Overall, an estimated 450,000 people are living with the disease in Thailand and an estimated 20,000 people die from it annually.

UNIVERSAL HEALTHCARE

Until the year 2000 almost one-third of all Thais had no healthcare coverage. In 2002, the **Universal Coverage Scheme or UCS** (popularly known as "The 30 Baht Health Scheme") was launched under the National Health Security Act. The clever nickname derived from the fact that patients had to only pay 30 baht for administration fees, no matter the prescription or operation, for each visit or admission to a hospital or clinic (in 2007, the co-payment was abolished and the UCS became free). The new card consolidated the Low Income Health Care Scheme and the Voluntary Health Card plan and incorporated 30 percent of the uninsured into the UCS.

Since the plan was hatched, almost the entire population has been covered by one of the three big health insurance policies: the Civil Servant Medical Benefit Scheme for civil servants (7.82 percent), the Social Security Scheme for company employees (16.6 percent), and the UCS for the rest of the primarily rural populace. The latter scheme also strove for a more egalitarian approach to public health, serving both the needs of the poor, who could not afford treatments, and helping those with means with costly treatments like chemotherapy for cancer. Thanks to the UCS, the household costs for taking care of family members stricken with catastrophic illnesses have been steadily declining, from a total of 5.7 percent in 2000 to only 3.3 percent in 2009.

However, one of the most severe side effects of the triple-pronged plan is that the number of outpatient visits has spiked, thereby putting strain on the system that taxpayers must also shoulder. To cope with that increase, the government has had to boost its spending to support these programs from 56 percent of total health expenditures in 2001 to 75 percent in 2010. For 2015, the total budget for the UCS

Strengths and Weaknesses of the Thai Healthcare System

STRENGTHS

> Thailand already achieved most of its health-related Millennium Development Goals (MDGs) by 2004, well in advance of the 2015 deadline. This prompted the country to pursue an MDG-plus strategy in recent years, with expanded targets like further reducing the rates of maternal mortality, HIV infections and malaria at regional levels, especially among the northern hill tribes and southern Muslim communities.

> The UCS covers almost the entire population.

> Strong technical capacities ensure that healthcare workers meet a high standard.

> The National Health Assembly (NHA) is an effective platform for developing policies through the exchange of scientific data and knowledge transfers with different sectors.

> The UCS also covers antiretroviral treatments for people living with HIV/AIDS, and has reduced the number of new HIV infections by more than 80 percent. It also fully covers dialysis and kidney transplants for patients with chronic renal failure.

> Bangkok's best private hospitals are so highly regarded and relatively affordable that they receive so-called "medical tourists" from around the world.

> The monarchy has a long-standing tradition of promoting nutrition, developing health projects and mobilizing funds to support public health.

WEAKNESSES

> Increased focus on profits by private hospitals and doctors, who can earn more by working in speciality areas or by focusing on medical tourists.

> A lack of focus on prevention of illnesses and chronic diseases through family medicine.

> There is inefficient management and integration of medical information.

> Spiritual well-being and traditional treatments have been pushed out of the mainstream with high-tech testing and equipment emphasized instead.

> Doctor-patient relationships have deteriorated with doctors either too rushed or too profit-driven to communicate with and assess patients effectively.

> There is a shortage and lack of even distribution across different regions of health resources (such as hospital beds) and personnel (such as nurses, dentists and pharmacists), especially in the more remote areas.

> A significant difference remains between the services and equipment of expensive private hospitals and that of cheaper but overcrowded public facilities.

> The discrepancy between the three schemes (the CSMBS, SSS and UCS) is a major source of inequity. Also, the CSMBS is inefficient as it pays on a fees-for-service basis. The overuse of medicine and diagnostics means that it costs four times more per capita than the UCS.

Universal Coverage Scheme (UCS)

This term is used to define a healthcare system that ensures all people obtain the treatments and services they need without suffering any undue financial hardship to pay for them.

amounted to almost 143 billion baht to cover around 49 million Thais, inclusive of anti-retroviral medicines for HIV/AIDS patients and renal treatments like dialysis.

The UCS is far from perfect. There is a disparity between the facilities in metropolitan areas and the countryside, just as there is a gulf between how the poor, the middle class and the wealthy are treated by medical staff — and Thailand needs to properly address these issues in order to fully achieve Goal 3.

That said, in tandem with the UCS has come a new way of looking at health. The existing system has generally been passive: when people get sick they go to the hospital. However, in the new millennium a way of thinking that advocates prevention has entered the public discourse and the mindsets of policy makers. Since then, "health" has been redefined in a broader context, covering physical, mental, social and even spiritual well-being. No longer just a matter for doctors and patients, health is starting to be recognized as an intrinsic part of human and social development as well as a fundamental right. With such strong medical institutions already in place, embracing this kind of proactive, holistic philosophy across the board would serve Thailand well in its efforts to promote healthy lifestyles and inclusive, equitable care.

TIME TO GET HEALTHY, THAILAND

In spite of Thailand's many victories on the health front, some uphill battles remain. In particular, the country needs to work harder to reduce the rate of teenage pregnancy and number of traffic deaths, and to promote healthy lifestyles among a population that is short on free time and increasingly engrossed in sedentary activities. Obesity has emerged as a growing public health threat and, put simply, Thais need to exercise more, eat healthier and get away from their smart devices once in awhile.

Indeed, with an overall obesity prevalence of 32.2 percent (as of 2011), Thailand ranks second in the ASEAN region — behind Malaysia's 44 percent — with the highest number of obese citizens, according to the World Health Organization. But it's not just adults. The 2012 MICS study found that 10.9 percent of children under age five are overweight in Thailand.

The key catalysts for Thailand's bursting beltlines are a decrease in physical activity and heavier consumption of processed foods, sugar and soft drinks. It may come as no surprise that Bangkokians are the most at risk of becoming obese. In today's fast-paced Thai society, many adults work six days a week and have long commutes. As a result, parents have less time to prepare meals at home and are relying more on fast food options and convenience stores like 7-Eleven to feed themselves and their kids. And the vast majority of what's on offer — processed meats, synthetic foods, boxed fruit juices, microwavable meals, candy, etc, — contain excess sugar, high fructose corn syrup, trans-fats and sodium. Far from nutritious, the lack of a balanced diet means that children aren't getting the nutrients they need to fully develop. A recent study by the Southeast Asia Nutrition Survey concluded that due to a lack of exercise and poor nutrition, in the next decade an increasing number of Thai children may be overweight, shorter and have lower IQs.

While Thailand is globally renowned for its cuisine, these days many of the staple dishes consumed by Thais aren't all that healthy either. Many are fried and heavy on oil. Fish sauce, shrimp paste and curry paste, which all figure prominently in Thai food, are all high in sodium. Just a half cup of coconut milk — a key ingredient in many soups, curries and desserts — contains over 200 calories and more than a day's recommended allowance of saturated fat.

Last but certainly not least, there's the sugar predicament. It would not be unfair to say that for a typical Thai person, their Achilles

> "With an overall obesity prevalence of 32.2 percent, Thailand ranks second in the ASEAN region in terms of the number of obese citizens."

OVERWEIGHT POPULATIONS IN SOUTHEAST ASIA

Overweight prevalence (%) for adults of both sexes (BMO of > 25kg/m2)

Country	%	Country	%	Country	%
Vietnam	10.2	Myanmar	18.4	Singapore	30.2
Cambodia	12.1	Indonesia	21	Thailand	32.2
Laos	13.3	Philippines	26.5	Malaysia	44.2

Source: WHO Noncommunicable Diseases Country Profiles, 2011

Health and Well-being Challenges in the Kingdom

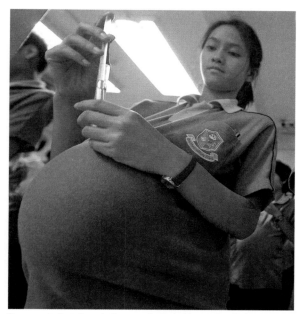

Teenage students have balloons under their shirts so they look pregnant during a sex education program.

SOCIETY

Teenage pregnancies Thailand's teenage pregnancy rate is the highest in Southeast Asia after Laos. Of every 1,000 teens aged 15 to 19, about 60 are mothers in Thailand – one of the highest rates in the world and more than 20 times higher than Singapore's, according to the SDG Index.

Aging society Around eight million people (13 percent of the population) are over age 60 in Thailand. With a younger generation that is less equipped and less interested in taking care of them, this puts added strain on both families and the healthcare system.

Smoke signals Despite more social awareness about the dangers of smoking, around 21.7 percent of Thais are still lighting up regularly as of 2014, down from 32 percent in 1991. Most are male, though there has been a slight increase in young women taking up the habit. The overall rise is attributed by some to the efforts of local and international tobacco industries counteracting the government's control measures through marketing strategies that violate the law.

Sedentary lifestyles With more and more Thais working in offices and spending their idle time parked in front of computers or using their smartphones, fast food and inactivity have become the prime culprits in a gargantuan increase in waistlines. More than one-third of Thais aged above 15 are fighting obesity today, which has also led to an upsurge in diabetes.

Road fatalities Thailand has the dubious distinction of ranking second in global traffic fatalities, with an estimated 24,237 road fatalities in 2012 and a mortality rate of 36.2 per 100,000, according to the World Health Organization.

GOVERNANCE AND TREATMENT

Bangkok magnet The concentration of health personnel in Bangkok and other urban areas in the central region remains high, especially in the private sector, leaving other areas like the Northeast region lacking in qualified staff. Physician density per 1,000 people was only 0.4 across the country, according to the SDG Index.

Overmedicating About 38,000 Thais die from antibiotic-resistant bacteria annually because of over-prescription and the prevalence of antibiotics in food and the country's water supply, according to research conducted by Chulalongkorn University's Drug System Monitoring Mechanism Development Centre.

Substance abuse In 2014 Thailand ranked first among ASEAN countries in alcohol consumption. Meanwhile, around 1.4 percent of Thais are addicted to amphetamine-type stimulants such as *yaba*, one of the highest rates in the world.

Mental health Only about one million Thais undergo regular mental health treatment, while the Department of Mental Health estimates that as many as 20 percent of Thais suffer from some form of mental illness including psychosis, anxiety disorders and depression.

Food security Feeding a growing population poses many problems, such as the quality and quantity of food, widespread access to nutritious food sources, the proper use of such resources and their long-term sustainability.

Good governance As with many parts of the public and private sectors, transparency and good governance are perennial challenges for the Ministry of Health.

heel is actually their "sweet tooth". According to Thailand's Ministry of Health, the average Thai citizen consumes 30 kilograms of sugar per year, which amounts to over three times the maximum recommended intake of 25,000 milligrams per day. These days, even in traditional cooking a hefty dose of refined sugar is added to the famous *pad thai* where previously cooks would have opted for tamarind paste as a sweetener. Not to mention the sugar-laced syrups and condensed milks heaped upon desserts and into coffee drinks and shakes.

And all that sugar consumption has consequences. According to the WHO, adults who consume less sugar have lower body weights, while evidence shows that increasing the amount of sugar in a person's diet is associated with an increase in weight. Excess sugar consumption can also lead to metabolic

THAILAND RANKED FIRST AMONG ASEAN COUNTRIES IN *Alcohol Consumption* IN 2014

Village Doctors

In the early 1980s, the number of doctors in rural areas remained insufficient, with one doctor per every 14,742 people in the South, one per every 12,419 people in the North and one per every 28,424 in the Northeast. To help villagers help themselves, in 1982 King Bhumibol launched the "village doctors" program, in which some villagers would receive basic healthcare training.

This idea was in many ways similar to the community health worker (CHW) idea that had gained currency in 1978 at the International Conference on Primary Health Care held at Alma-Ata in what is today Kazakhstan. For decades, some developing countries had been instructing and using grassroots-level workers, called extension workers, in agricultural and community development programs. They in turn trained people in their communities, spreading knowledge and development.

The Ministry of Public Health had experimented on a very small scale with community health workers, but the security situation in the late 1970s and some opposition from medical professionals inhibited its development. The king's project was launched with volunteers in Chiang Mai province. Selected villagers were taught prevention techniques, how to administer first aid and about nutrition – especially in relation to pregnant mothers and young children. The use of traditional herbal medicines for simple ailments was encouraged, but the "village doctor" would also learn when and how to refer patients to district or provincial hospitals for more serious treatment. The training was later expanded to Sakon Nakhon and Narathiwat, both provinces where King Bhumibol maintained a residence. Thousands of village doctors across the country were eventually trained.

At first, as with the Ministry of Public Health's attempts at a similar initiative, there was resistance. "A group of doctors was against it. Their reaction was that these people are not qualified to be delivering healthcare. But what they are delivering is very basic," said Dr. Sumet Tantivejkul, secretary-general of the Chaipattana Foundation. Resistance soon gave way to support, especially as the government of Prime Minister Prem Tinsulanonda launched its own rural development program. Prem, who had once been regional army commander for the Northeast, supported the king's belief that development could defeat insurgency, and his successive administrations promoted the king's strategies.

Around the same time the Ministry of Public Health chose to scale up its own community health worker program. In the ministry's version, the doctors are community health volunteers, and by the turn of the millennium more than one million of them had been trained nationwide.

syndrome (pre-diabetes) and even Type 2 diabetes, a growing problem in the kingdom.

But Thailand's obesity problem isn't just about diet. The rise of personal smart phones, laptops and tablets mean that fewer children and young adults are using their spare time for physical activities like recreational sports or hitting the gym. The prevalence of cheap taxis, motorcycle taxis, public transport — and the fact that so many people own their own cars and motorbikes — also means that fewer Thais are willing to walk from Point A to Point B, even for short distances. In Bangkok, the lack of green spaces and public parks doesn't help matters either.

The country's Health Department launched a "No Belly Fat" campaign in 2014, urging people to eat right and exercise, and the Thai Health Promotion Foundation has since 2001 campaigned on numerous health issues in the kingdom. But while health agencies can supply the information you need and gyms can provide the exercise equipment, the choice of whether or not to eat right and lead a healthy lifestyle ultimately falls on the individual.

THE ONGOING STRUGGLE TO CURB SUBSTANCE ABUSE

Alcoholism, narcotic drug abuse and smoking also remain substantial health threats. Of particular concern is addiction among the country's youth.

Thailand ranked third of ASEAN countries in terms of youth smoking in 2014, according to a study by the Tobacco Control Research and Knowledge Management Centre (TRC) of Mahidol University. Recognizing the need for stronger anti-smoking legislation, the Thai cabinet is drafting legislation that would raise the minimum age to purchase cigarettes from

A woman has her blood pressure checked at an elderly day care center in Bangkok.

18 to 20. The law would also outlaw the sale of individual cigarettes, prohibit cigarette sales in some public places, outline stricter punishments for violators of the law, and ban cigarette companies from advertising their products as sponsors of contests and social events. Despite more social awareness about the dangers of smoking, around 21.7 percent of Thais are still lighting up regularly as of 2014, down from 32 percent in 1991.

In recent years, Thailand has also taken a number of steps to try and curb alcohol consumption. Laws are enforced nationwide to limit the hours during which alcohol can be sold, and in 2015 the Alcohol Control Act was amended to prohibit the sale of alcoholic beverages within a 300-meter radius of higher-education institutions. Alcohol imports are also taxed heavily, though domestic products remain inexpensive. Despite the implementation of these measures, in 2014 Thailand ranked first among ASEAN countries in alcohol consumption, according to the WHO.

In terms of treatment, Thailand still needs to scale up efforts to encourage alcoholics to seek help. The social stigma associated with

Pollution and Toxic Chemicals

Reducing pollution and improving safety education on the proper handling of hazardous chemicals remain two areas where Thailand needs to improve on its path toward achieving Goal 3 by 2030.

In the kingdom's agriculture sector, the reliance on hazardous chemicals has been linked to a number of health afflictions including cancer, and developmental and reproductive diseases. Many toxic substances like DDT, which was banned in the United States in 1972, are still widely used by Thai farmers. Lacking the proper schooling on application, Thai farmers have often over-used pesticides and over-fertilized their crops. Indeed, Thailand's heavy use of agro-chemicals has increased dramatically in recent years, as imports rose from 78,000 tons in 2005 to 160,000 tons in 2011.

But such risks are not just confined to the fields. The variety of accidents involving improper handling of chemicals in Thailand suggests an overall need to improve safety education.

For example, in March 2016, eight people suffocated to death at Siam Commercial Bank's headquarters in Bangkok when workers accidentally released Pyrogen aerosol, a chemical fire retardant normally intended to deprive fire of oxygen. In recent years, a number of foreign tourists are also thought to have died after exposure to toxic chemicals at their hotels.

Air and water pollution, as well as the disposal of waste, have also been major health concerns for decades. The economy's increasing reliance on manufacturing and industry means that more people are living in close proximity to factories and dumpsites. In both March 2014 and April 2015, massive fires at a dumpsite in Samut Prakan province forced the evacuation of hundreds of residents living in communities nearby.

Meanwhile, in the northern provinces the burning issue is the smoky haze that descends during the hot season every year, which blurs vision, clogs up lungs and causes fits of coughing that leads, in extreme cases, to terminal diseases like lung cancer. In 2016, tests conducted by Chiang Mai University found that the amount of small dust particulates in the air had significantly exceeded the safety limit of 120 micrograms per cubic meter, ranging from between 131 to 163 micrograms per cubic meter in some areas.

alcoholism means that few people want to admit openly that they have a problem. As a result, few of those suffering from alcohol addiction attend treatment programs, even though government-sponsored programs are free of charge.

The kingdom has also long struggled to contain rising drug abuse. Around 1.4 percent of Thais are addicted to amphetamine-type stimulants, among the highest percentages in the world, according to the United Nations Office of Drugs and Crime. Between 2009 and 2014, the number of drug cases in Thailand more than doubled from roughly 151,000 to about 347,000, while the number of registered drug offenders increased from nearly 168,000 to almost 366,000, according to official figures. As manufacturing and trafficking from Myanmar continues to increase, methamphetamines (known locally as *yaba* in its pill form) have become cheaper and easier to obtain in Thailand. In fact, a single *yaba* pill now costs as little as US$2.

Admitting that Thailand's fight against *yaba* is failing, in July 2016 Justice Minister Paiboon Koomchaya proposed removing methamphetamines from the dangerous illicit narcotics list. Defending the notion, the minister reasoned that efforts to crack down on narcotics had led to the arrests of traffickers and users in such huge numbers that jails were overcrowded, while authorities had failed to make significant headway on arresting kingpins.

Meanwhile, the spread of HIV among injection drug users is higher than in all other vulnerable populations in Thailand. Among the kingdom's estimated 30,000 injection drug users, there is an HIV prevalence of almost 30 percent among women and 24.5 percent of males, according to a 2014 UNAIDS report. This high prevalence boils down to the fact that by and large, harm reduction services such as needle and syringe exchange programs still aren't widely available. In 2012 it was reported that only 12 sets of clean needles and syringes were distributed per person who injects drugs per year in Thailand (against a national target of 88 sets and the regional median of 116 sets), according to the Thai National AIDS Committee's "2014 Thailand AIDS Response Progress Report".

Civil society-sponsored drop-in centers and other treatment initiatives have sought to fill these gaps over the years, but have not been able to pin down the kind of long-term funding necessary to make such programs successful, or sustainable. While law enforcement agencies have historically taken an approach focused primarily on "prohibition, punishment and forced rehabilitation", analysts like Pascal Tanguay, formerly of the Asian Harm Reduction Network, argue that government support of civil society-led initiatives to scale up harm reduction and treatment for injection drug users would yield better results. Funding for such programs remains a key challenge, though sources like UNAIDS and the Global Fund to Fight AIDS, Tuberculosis and Malaria have provided the kingdom with significant resources in the past.

Across the board, where addiction and substance abuse are concerned, the key challenge for Thailand lies in implementing proactive measures that combat these societal scourges while avoiding practices that further stigmatize already alienated groups.

THE PRUDENT APPROACH

While the two-tiered system of expensive private versus inefficient public healthcare dominates the dynamic in Thailand, there are examples of efforts that present a more balanced and prevention-oriented way of care. The Thai Health Promotion, which was established in 2001, aims to encourage, motivate, support and cooperate with other agencies to enhance the overall well-being of people in Thailand. It focuses on the physical, mental, and social aspects of life, encouraging the public to make smart and healthy choices. Through awareness campaigns and other methods, its main thrust is thus prevention.

Meanwhile, Theptarin Hospital, a small private hospital that focuses on the treatment of diabetes, is following the principles of SEP. It also focuses on prevention. Instead of pursuing short-term profit schemes and highly specialized procedures like the bigger hospitals, Theptarin Hospital devises strategies to help patients engage in self-care through education and better diets, for example.

Theptarin Hospital also emphasizes developing the virtuous behavior of its personnel by promoting from within, encouraging teamwork rather than individual success and highlighting the importance of strong ethics. Thus the staff at Theptarin have created a more community-oriented culture that is focused more on the mission of care than the creation of profits.

Students meditate at a school in Surin. A focus on spiritual well-being is returning to Thai schools.

Further Reading

• *The Kingdom of Thailand Health System Review* by Patama Vapatanavong and others, the Asia Pacific Observatory on Health Systems and Policies, 2015

• *Thai Health Report* by the Institute for Population and Social Research, Mahidol University, 2013–2014

• *ePatient 2015: 15 Surprising Trends Changing Health Care* by Rohit Bhargava and Fard Johnmar, 2015

4 QUALITY EDUCATION \ Ensure inclusive and equitable quality education and promote lifelong learning opportunities for all.

QUALITY EDUCATION
Reform in the Name of 21st-Century Skills and Challenges

Thailand has been integrating Sufficiency Economy Philosophy into its national curriculum for about eight years.

Calls to Action

- Promote student-centered learning and the development of critical thinking skills; phase out the use of rote learning techniques
- Guarantee equal access to quality education and skills training in order to produce more qualified graduates and eliminate illiteracy
- Encourage teachers to be more creative and innovative; improve the quality of teacher training programs
- Expand integration of SEP into more Thai schools as a way to help ensure that all students acquire the knowledge and skills needed to promote sustainable development and become life-long learners
- Address the deep urban-rural educational divide
- Promote accountability and stronger implementation of education reform

Thais believe school is a second home for their children. Many parents, however, think that this "home" is in drastic need of serious renovations. And parents are not alone in their concerns. Thailand does not receive high scores in global education rankings, and studies pinpoint the need for further reforms. Indeed, the country is in the middle of a second round of education reforms that runs from 2009 to 2018. The first began with the National Education Act (NEA) in 1999, which followed on the heels of the financial crisis and a new constitution. It boosted enrollments in both schools and universities, and also guaranteed an education for those suffering from disabilities or social deprivation. The second round is focused much more on improving the quality of Thailand's education system. Thailand's relatively strong score on the SDG Index does not make up for the fact that Thailand should do far more to better prepare its students.

These misgivings are borne out by other measures — both internal and global — that show Thailand's education system is performing too low in the rankings, particularly in core subjects such as math, sciences and English. In the 2016–2017 Global Competitiveness Report of the World Economic Forum, Thailand's primary education ranked seventh and its higher education fourth out of the nine ASEAN countries included in the report (no data was available for Myanmar). In the latest ranking from 2012

> "You should not think that you are studying in order to pass an entrance exam, because our existence doesn't depend on whether you marked down the right multiple choice answer on your exam. Our existence depends on working and analyzing various problems."
> -King Bhumibol Adulyadej

of the Programme for International Student Assessment (PISA), Thailand placed 50th out of the 65 nations that participate in this measure of the performance of 15 year olds in reading, mathematics and science. The PISA study also determined that Thailand had the sixth-largest gap in the quality of educational resources between socioeconomically advantaged and disadvantaged schools among the 62 countries included in this metric.

According to the Times Higher Education University Rankings of 2015–2016, Thailand had only two universities among the top 100 in Asia, and just one (Mahidol University) in the world's top 600. These rankings cannot be explained away by a lack of financial resources. Thailand has consistently allocated about 20 percent of its national budget — about four percent of annual output — to education, which is among the highest totals in the world. Much of the increased spending in the last decade has addressed the poor pay of public school teachers, once regarded as the system's main drawback. In 2014, the starting salary of a teacher was raised to 15,050 baht per month, up from 7,630–8,340 baht per month during 2009 to 2012.

That said, there are certainly bright spots and Thailand has already achieved many of the specific benchmarks outlined in Goal 4. Most notably, there's near universal access to free education and vocational training, and by and large there are few gender disparities. Thus, girls have the same access as boys to early childhood development, care and all levels of education and training. The kingdom has high literacy rates among both adults and children. It's proactive in incorporating sustainable development principles into curricula. Schools are safe and non-violent. And, the vast majority of education facilities are also child-, disability- and gender-sensitive.

However, parents still feel that their children are not getting enough out of their well-financed school system. One obvious manifestation is the high sums parents spend on some 5,000 "cram schools" across the country that provide supplementary education for all ages. Thai students spend hours upon hours each academic year in this kind of intensive, supervised studying environment. Despite their popularity, most of these tutoring centers focus primarily on memorization to prep students for standardized exams. Some experts say that this kind of cramming is indicative of the main shortcoming in Thailand's education system: the lack of **student-centered learning.**

While the student-centered approach has been hailed around the world as the way forward, in Thailand such techniques are still shunned in favor of "rote learning" — a grinding and stultifying process of drilling, memorizing and copying that leads to passive students who lack critical thinking skills and have trouble competing in a world where innovation and initiative matter most.

Student-centered learning
The principle of creating a learning environment designed to encourage children to discover new skills and knowledge, with teachers facilitating rather than providing front-of-the-classroom teaching.

It's not that the authorities are unaware of the issue. As far back as the year 2000, the country officially adopted student-centered learning, at least in theory. However, more than 15 years later, child-centered learning is still a rarity in most Thai schools.

According to Thitinan Pongsudhirak, a political scientist at Chulalongkorn University, "The mindset is from the nation-building and Cold War period to produce obedient and nationalistic citizens, which does not fit 21st-century needs. It is hierarchical, top-down, with a systematic lack of critical thinking." The education system's rather dated emphasis on conformity rather than individuality has also been called into question.

For example, Thailand is one of the few countries in the world that still requires even university students to don uniforms. Meanwhile, at the primary and secondary level there are strict hair length guidelines that are often enforced with the canceling of an afternoon's classes and a pair of clippers in the hands of the school's Head of Discipline. In many ways, this insistence on students looking the part versus students who can actually think for themselves hints at much deeper-seated problems.

Related to this is the nation's dated assessment system. The multiple-choice **Ordinary National Educational Test (O-NET)** and Advanced National Educational Test (A-NET) run by the National Institute of Education Testing Services (NIETS) have been roundly criticized, mostly because they evaluate memory rather than thinking, but also because of the narrow, parochial and moralistic nature of some questions and the "correct" answers.

Far from being just a small part of the problem, such exams also influence teaching styles, reinforcing the tendency of teachers to fall back on rote learning. This combination of memorization-based exams and a lecturing style of instruction tend to produce passive, uncreative students. Furthermore, Thai students are almost never failed and made to repeat a grade – no matter how bad their academic performance.

When examining the Thai education system, the onus has often fallen on the teachers, who, to be fair, frequently fail to impress. In 2010, the Office of the Basic Education Commission (OBEC) conducted the first national exam to test secondary teachers in their own subjects. The result was that 88 percent of computer science teachers, 86 percent of biology teachers, 84 percent of math teachers, 71 percent of physics teachers and 64 percent of chemistry teachers failed their own subjects.

Thus, while the country has met the UN's Millennium Development Goal of universal primary education for all, the UNDP notes that "the greatest challenge lies in improving quality." To achieve Goal 4, the kingdom needs to embolden its teachers to be more creative, innovative and engaging, and amp up the quality of teacher training initiatives.

Ordinary National Educational Test (O-NET)

A standardized test on eight major subjects administered by the National Institute of Educational Testing Service (NIETS) to assess the academic proficiency of 6th, 9th and 12th graders in Thailand.

Thailand Offers 15 Years OF FREE EDUCATION FOR ALL

Some 35,000 students sit for an exam hoping to gain admission to Srinakharinwirot University.

Accountability is also key. As TDRI's Somkiat Tangkitvanich and Supanut Sasiwuttiwat argued: "Effective education reform must incorporate the creation of an accountability system." This, they say, is vital to the success of other reforms. Such a system should include improved standardized tests and report cards, rewarding teachers and administrators according to student performance, and providing "demand-side" financing. It's also high time for the book on rote learning to be shelved for good.

AN OVERHAUL IN THE MAKING

With the coming to power of the National Council for Peace and Order under General Prayuth Chan-ocha, some educationalists had hoped the focus on reforming all of Thailand's institutions might rejuvenate the effort to transform the country's education system. However, so far the emphasis has been mostly on the moral side.

In June 2014, in his nightly address to the nation, the general who became prime minister said the Ministry of Education should include in the curriculum subjects that help reinforce the values of "Being Thai": national pride and upholding the institution of the monarchy. "The purpose is to instill discipline, strengthen the physical and mental state, and reinforce conscience and social responsibility", Prayuth said.

Almost exactly two years later, in June 2016, Prayuth announced that free basic education was being expanded from 12 to 15 years in order to cover kindergarten levels as well primary and secondary. The new 15-year state-subsidized plan provides free education through the Mathayom 6 (Grade 12) level, or the equivalent in vocational schools. It also covers special education for students with disabilities or special needs, and "welfare education" for underprivileged children.

Students learn vocational skills like weaving at a school in Lam Plai Mat district, Buriram province.

21st-century skills

A range of skills, knowledge, habits and character traits necessary to compete in the modern world. They include critical thinking, problem solving, synthesizing information, research skills, creativity, self-direction, innovation, digital skills, financial literacy and civic/ethical literacy.

Due to the country's chronic shortage of skilled workers, subsequent governments over the past decade have tried to enhance and expand vocational education and to encourage more students to enroll in these classes. For its part, the current government increased the budget of the Office of the Vocational Education Commission (OVEC) by about a third in 2015. The OVEC has also developed a new curriculum at a higher-level certificate of vocational education to teach more **21st-century skills** and to meet the challenges of the ASEAN Economic Community (AEC), which came into being at the end of 2015.

Further reforms of the education system are to be overseen by a new "super board" that was established in 2015. According to Amornwich Nakornthap, the spokesman of the National Reform Council's committee on education and human resource development, the committee has been drafting a law to decentralize power to local authorities. In theory, this should help schools and local authorities design and manage an education plan that better suits their environment, including their budgets.

It's worth noting that there has already been some progress in improving governance and decentralization of the education system. The global Programme for International Student Assessment (PISA, 2012) found that Thai schools were comparatively more responsible for curricula and assessments than schools in other participating countries (Thailand ranked second out of 64). The percentage of students in schools whose principals and/or teachers have considerable responsibility in determining course content was also one of the highest (ranking seventh out of 64).

THAILAND'S HEAD START ON SUSTAINABLE EDUCATION

Despite the overall shortcomings of Thailand's education system, one area in which the kingdom has proven to be a trend-setter is in the promotion of sustainability in schools. Some eight years of applying the principles of King Bhumibol's Sufficiency Economy Philosophy (SEP) in the national education system have yielded positive results overall. Likewise, the goal of SEP within education reform, which is to promote sustainability practices and theories as a basis for national development, is in line with Target 4.7, which aims at ensuring that "all learners acquire the knowledge and skills needed to promote sustainable development."

To whit, SEP offers a model for responsible behavior through "sufficiency thinking" based on three principles: moderation, reasonableness and prudence. Another component of SEP is the understanding that all activities and decisions should be undertaken while striving to remain virtuous, and that individuals should apply any and all relevant knowledge at one's disposal. Furthermore, SEP encourages what is called "progress with balance in life's four dimensions": the economic, environmental, social and cultural aspects of life. Because of the holistic nature of this approach, advocates of SEP say that it impacts the "head, heart and hands" of students – the intellectual, spiritual and practical aspects of education.

The overall aim of the philosophy is to instill sufficiency-based attitudes and practices that help students form lifelong habits of thinking and doing that ultimately support the building of a sustainable society. Students are encouraged to develop the kind of mentality and skill sets that foster a self-reliant, balanced lifestyle. Furthermore, SEP-based curriculum is designed to inculcate a disciplined, moral and ethical outlook that reflects virtue and honesty.

In SEP schools, learning through the process of doing (questioning, planning, acting and reflecting), utilizing collective local and global knowledge, and developing decision-making skills based on SEP are essential. Students are taught to use reasoning, prudence and carefulness when applying knowledge, and to do so with the aim of contributing to the betterment of their schools and communities.

In the early stages of SEP implementation, schools used a curriculum intended to cultivate SEP-based behavioral principles in students. However, it was determined that classroom

> "Teachers and principals should be made more accountable to students and parents by linking their remuneration to improvements in students' learning outcomes. This should be coupled with enhancing 21st-century skills among Thailand's next generation."
>
> -Dr Somkiat Tangkitvanich, President of Thailand Development Research Institute

Pupils take part in an activity at a Sufficiency Economy certified school.

Challenges Facing the Thai Education System

HUGE BUDGET Thailand spends more of its national budget on public education (20 percent) than any other item. This is relatively high by global standards, yet the quality of education does not meet those standards. Such a poor return on investment is probably the country's biggest challenge in this area.

SLOW TO EVOLVE Despite frequent announcements of education reform over the years, including a desire to shift to more child-centered learning methods, little has been achieved because of bureaucratic inertia, weak implementation, lack of accountability and policy discontinuity resulting from unstable politics. A highly centralized, bureaucratic national testing system reinforces rote learning and standardization. Pass rates are even substantially lower than those for global testing schemes, which suggests that there is something amiss with the tests themselves rather than the students taking them.

TEACHING CREATIVITY At times, inadequately qualified teachers wedded to traditional top-down teaching methods preside over passive pupils who are discouraged from questioning or thinking for themselves. Without more creative and innovative graduates, Thailand is likely to remain trapped in the middle-income zone, unable to meet the challenges of an increasingly competitive and global "knowledge economy".

LEARNING MORE ENGLISH English proficiency in Thailand is near the bottom of ASEAN's class and shows no signs of improving, a serious impediment to Thailand's competitiveness as international business becomes increasingly integrated.

VOCATIONAL TRAINING Successive attempts over the past decade to promote and improve the quality of vocational training at upper secondary and tertiary levels have delivered weak outcomes leading to an undersupply of skilled vocational workers. One reason for this is that high school graduates are increasingly opting to attend universities instead of vocational institutions.

UNQUALIFIED GRADUATES Despite higher enrollments, Thai universities are still turning out graduates in social sciences and the arts who struggle to find jobs. Meanwhile, the sciences are being relatively neglected to the detriment of employers who cannot fill vacancies that require such skills. In 2011, UNESCO reported that Thais with tertiary qualifications lacked the following: "communication skills, computer and ICT-using abilities, management, calculation skills, problem solving, teamwork, responsibility, honesty, tolerance, discipline, punctuality and leadership."

LEARNING FOR LIFE As Thailand grapples with reforms in 11 areas of public life, education is in danger of being elbowed out by other issues. Consequently, educationalists fear that curriculum changes intended to improve public morals and military-like rigor are taking precedence over the kind of reforms necessary to promote critical thinking and life-long learning skills.

ACCOUNTABILITY Thailand needs more accountability on both sides of the teacher's desk. Experts argue that remuneration for teachers and headmasters should be linked to improvements in students' learning outcomes. Meanwhile, students should be held to higher standards, disciplined for cheating and plagiarism, and made to repeat a grade if warranted by an inadequate academic performance.

DEEP DIVIDE Education provisions and outcomes are still very unequal across the country, exacerbating the long-standing urban–rural divide. Meanwhile, expensive and elitist private and international schools turn out better-qualified young people disconnected from the majority of their compatriots.

GREYING TEACHERS The country's aging demographic means that about 40 percent of the current total number of teachers will retire in the next ten years. Replacing them with high-quality graduates will prove difficult, especially outside urban areas.

RETENTION RATE Only 75 percent of children are entering Grade 1 at age six in Thailand. In the Northeast, that figure dips to 65 percent. The retention rate of students over the 12 years of basic education has also been declining. Only 60.7 percent of those who entered Grade 1 in 2001 were still in school in 2012.

teaching alone was not enough to alter students' perspectives and actions, and so a "whole school" approach was developed. This approach applies sufficiency thinking in all school activities including learning systems, extracurricular student activities, school management and community relations.

Furthermore teachers, school administrators, parents and community members are groomed to act as SEP role models, creating a school and community environment that helps to foster sufficiency mindsets in students.

Two levels of school certification for sufficiency-based schools have been established by the Ministry of Education. The first certification level for sufficiency-based schools concerns an SEP curriculum and a whole-school approach. By April 2016, of the approxi-

UNESCO's Education for Sustainable Development Campaign

Education for Sustainable Development (ESD) is a UNESCO global campaign that reflects its vision for a world where everyone can benefit from learning the values, behavior and lifestyles required for a sustainable future. ESD takes a holistic approach that identifies environmental sustainability with the sustainability of society. It is to be promoted both informally and through all educational levels. The aim is to build capacity for community-based decision-making, social tolerance, environmental stewardship, an adaptable workforce and improved quality of life for all, using techniques that promote participatory learning and informed thinking.

The key features of sustainable schools are listed here:

1. A commitment to positive social, environmental and economic outcomes, focusing on sociocultural dimensions of sustainability rather than a restricted focus on "green" agendas.
2. Visionary school leadership that encourages involvement and consensual decision-making.
3. Continuing professional development for teachers and all participants.
4. Extensive multi-stakeholder partnerships (i.e., community, government authorities, the private sector, school networks) that emphasize the active engagement of multiple actors in the joint redesign of basic operations, processes and relationships of school-related activities.
5. Curriculum committed to sustainability.
6. The school as a "learning organization", using participatory learning approaches for students and reflective practice for teachers. Sustainable education requires integrative, problem-based and exploratory forms of learning that invite participants to be critical, creative and change-oriented.
7. Whole-school approaches with sustainability practices in all aspects of school activities and everyone's lives. They coordinate sustainable learning activities between school and community.
8. Expertise in education for sustainability.
9. Appropriate political support (i.e., through the Ministry of Education and other governmental agencies).

Source: UNESCO

The SEP Approach

An interview with Priyanut Dharmapiya, Director of the Sufficiency School Center, Foundation of Virtuous Youth, Bangkok.

Why should Thailand instill a mindset in students based on King Bhumibol Adulyadej's Sufficiency Economy Philosophy (SEP)?
World citizens are facing shared problems regarding environmental deterioration, conflict and gradual decline in cultural heritage and spiritual values. There is unbalanced development among the social, environmental, cultural and economic or material dimensions of living. In Thailand, we have applied SEP principles bestowed by King Bhumibol as a systemic process to restore balance in developing the country. In order to be truly sustainable, we have to start with cultivating an SEP mindset in our younger generations so that they can have the knowledge and skills needed to build a sustainable society.

What are the successes so far?
The education sector has started to understand and apply SEP, with over 21,000 schools having been certified as "sufficiency-based schools". They are schools in a variety of contexts (e.g., in religion, size, geographical characteristics). We have gone through a learning process and have understood how to develop sufficiency-based schools. However, to ensure sustainability, we have to find ways to strengthen the SEP culture in these schools. Currently, 121 schools qualify as Sufficiency Education Learning Centers and have developed a very strong SEP culture, and are therefore able to coach and mentor other schools and educators.

What are the challenges?
We try to identify as many different models or building blocks as we need for developing sufficiency-based schools in different contexts. That will help in reducing the risks of trial and error when schools start implementing SEP. Furthermore, even though we cultivate a SEP mindset in students and their local communities, when students finish school and enter the wider society such as moving to other communities or provinces, their new surroundings might

> "World citizens are facing shared problems regarding environmental deterioration, conflict, and gradual decline in cultural heritage and spiritual values....In Thailand, we have applied SEP principles bestowed by King Bhumibol as a systemic process to restore balance in developing the country."

not be compatible with SEP thinking. Therefore, it might be difficult for them to sustain their SEP mindset and practices. The development of sustainability in Thai society has been slow partly due to a decade of countrywide political turmoil. We in the education sector can only start SEP with individual schools and surrounding communities. Success in this undertaking requires participation from other larger sectors of society. In fact, the society at large, including the media, are the students' teacher, often influencing student behavior. We need more role models from all sectors.

Can SEP be applied in other countries?
Yes. SEP shares many similarities with the main concepts of international approaches to developing sustainability, such as focusing on balanced development between economic, environmental, cultural and social matters. SEP is also in line with international character-building methods such as learning by doing, learning through reflection or self-discovery, rather than memorization – together with what we call "21st-century skills (e.g., higher-order thinking, creativity and good citizenship)." In addition, SEP has its own special contribution to make a more complete and holistic framework of decision-making mechanisms (prudence, integrity, reasonableness and moderation, for example), instead of focusing on just one or a few decision-making elements. It's all about character-development focused education which many countries are implementing.

The Value of Distance Learning

At the start of his reign, King Bhumibol Adulyadej had to face the stark fact that many of his subjects had no access to education outside of temples. From the start, the king focused on filling these gaps so that all Thais could receive some degree of formal schooling. In some cases, this simply meant opening new schools. But other initiatives took advantage of new technology to increase access to knowledge.

In 1995, His Majesty's interest in the potential offered by distance education led to the establishment of the Distance Learning Foundation and a flagship school for broadcasts at Klai Kangwol palace in Hua Hin. The project is particularly useful to schools without teachers to cover key subjects. To date, more than 35,200 schools have been equipped to receive the broadcasts nationwide. Through the country's leading cable television provider, the broadcasts can also be received in homes, or just about anywhere with a TV.

Broadcasts cover both primary and secondary school curricula, as well as vocational training, community education, university education and classes in six languages. Classrooms at Klai Kangwol School serve as studios for the telecasts, and in addition to teaching some 2,000 in-house students across 12 grades, faculty and staff also produce educational programming on 15 different channels that broadcast 24/7. Indeed, the project produces hundreds of hours of new educational programming each month with only 60 teachers and 110 technical staff. On the listening end, educators are expected to pull their weight and are provided with two-inch thick handbooks containing broadcast schedules, lesson plans, handout materials and homework assignments that can be easily photocopied.

While most broadcasts cover the national curriculum, daily programming on three channels and weekend programming on all channels focus on non-formal learning, including anything from cooking to foreign languages. Despite its overall success, the project has experienced a few hitches. In 2009, the Ministry of Education determined that many schools were not making good enough use of the educational broadcasts. In addition, a survey by the Office of the Basic Education Commission found that 3,964 satellite dishes and learning equipment for the project had gone unused. In response to these findings, the project distributed extra handbooks and set up a free telephone call-in system. It also experimented with teleconferencing to remote schools, and enlisted the National Electronics and Computer Center to make lessons available on the Internet.

The enduring problem of finding teachers willing to stay at remote schools means there will continue to be a need for distance learning, but technology will never offer all the answers. In 1996, soon after the start of the distance learning broadcasts, the king commented: "Nowadays, we have high technology that offers better means of expanding the schools and disseminating knowledge, but there is nothing that can replace an education that develops knowledge and edifies the mind."

---

mately 40,000 schools in Thailand, nearly 21,000 had been certified as having successfully integrated sufficiency economy as a practical orientation in all aspects of school activities and daily lives.

The second type of school certification in SEP is the **Sufficiency Education Learning Center (SELC)** for schools that can offer advice, mentoring and supervision to other schools that aim to become sufficiency accredited. By mid-2016, there were 121 accredited SELCs. In addition to the two formal certification levels, the Sufficiency School Center under the Foundation of Virtuous Youth has created a "best-practice" status that serves to help sufficiency-based schools at the initial certification level to improve their quality of learning activities, innovation, quality performance and management. Those selected for the best-practice status receive opportunities

Thai students learn about agriculture at a Sufficiency Economy accredited school.

Sufficiency Education Learning Centers (SELCs):

Special schools that are accredited to train and coach teachers and administrators of other schools that are striving for Sufficiency Education Philosophy (SEP) accreditation.

to participate in various SELC programs and training. As of June 2016, there were around 165 best-practice SEP schools.

For most government schools, the quest for SEP certification often begins with the aim of improving management of its limited resources. In contrast, few private schools have sought SEP accreditation. Those that have done so typically have an abundance of resources at their disposal and want to apply SEP as a way to promote moderation, sharing and good citizenship among their students, who primarily come from well-off families.

According to research conducted by the Sufficiency School Center of the Foundation of Virtuous Youth, students who attend SEP schools demonstrate greater moderation, a better ability to efficiently utilize and share limited resources, and are more likely to volunteer in their schools and communities. They participate in and are proud of their local cultural activities, and show enhanced analytical and social skills.

In addition, the research found that students from sufficiency schools demonstrated higher levels of proficiency than students in other schools in all five competencies, which include life skills, communication, logic, problem solving and IT literacy. In short, they are acquiring "21st-century skills".

While academic grades are never the primary focus of sufficiency schools, the study determined that scores of students from SEP schools on the O-NET exam tend to be higher than those of students from other schools.

According to the survey, parents have also noted the positive changes taking place in their children and in their communities. Where the SEP approach is taught, parents see schools and surrounding communities forging partnerships to identify and solve local problems. By working together, they can address issues that hit close to home and by incorporating SEP they can do so in a manner that is sustainable and equitable.

Further Reading

• *A Critical Study of Thailand's Higher Education Reforms: The Culture of Borrowing* by Rattana Lao, 2015

• *A Decade of Education Reform in Thailand: Broken Promise or Impossible Dream?* by Philip Hallinger and Moosung Lee, Cambridge Journal of Education, June 2011

• *Toward a Sustainable Society: Cultivating Sufficiency-Mindset in Thai Schools* by Priyanut Dharmapiya and Molraudee Saratun, 2016

5 GENDER EQUALITY

Achieve gender equality and empower all women and girls.

GENDER EQUALITY

Empowering 34 Million Thai Women and Girls to Meet Their Full Potential

Khamhla Tongburan at the Dignity Returns factory, which is run by a collective of former factory workers who aspire to establish their own worker-owned factory providing fair wages, fair working hours and a safe environment.

Calls to Action

- Address problems involving violence against women and girls, trafficking in persons, sexual exploitation, forced labor and forced panhandling
- Implement better labor protections, particularly for domestic workers due to the informal nature of the work and closed setting
- Reduce the violent objectification of women in the media and establish sex trade regulations that do not discriminate solely against sex workers
- Increase the rate of women taking decision-making level positions, particularly in parliament and the main government agencies
- Ensure women equal decision-making power in land or property ownership and control, as guaranteed by law

Gender equality is not just about making life better for women and girls. Of course, by achieving gender equality the lives of half the world's population, including more than 34 million females in Thailand, could be vastly improved. But gender equality is also about increasing the social, economic and environmental health of the world in general.

For example, in the private sector, gender equality can have a positive effect on productivity and thus lead to greater GDP. In the arena of reproductive health, promoting gender equality is an investment in the next generation. Healthier, better-educated mothers have healthier, better-educated children, which has a direct effect on the offspring's well-being and future prospects. In the public sector, strengthening women's voices can enhance the quality of developmental decision-making and contribute to the enactment of laws that provide more comprehensive legal protections. These types of advancements in gender equality are also closely tied to social justice for other minority groups. The protection of women's rights and those of the LGBTI community, which are deeply intertwined, can help minimize social prejudice and promote human rights among individuals in marginalized and vulnerable communities, such as ethnic or religious minorities, the disabled and the elderly.

When it comes to agriculture and the environment, women have an integral role to play as well. Women are responsible for over half of the world's food production overall and up to 60–80 percent in developing countries. On local levels, women are also influential in establishing sustainable use of resources, such as in small-scale fishing communities. In these roles, their extensive knowledge of natural resource and ecosystem management and their contribution to environmental sustainability should be taken into account. In addition, according to the United Nations Entity for Gender Equality and the Empowerment of Women, studies on natural disasters show that environmental health and the well-being of women are inextricably tied. During natural disasters, the recorded mortality rate among women is far higher than men. In some cases where natural disasters eliminated or destroyed main sources of income, incidents of trafficking surged. This appears to indicate that climate-related events will often reveal the vulnerability of women, and therefore climate change resilience and preparedness can help empower women to a dramatic degree.

> "Thai women still earn less than men in every major sector. Women are also far more likely to be employed part-time or on a temporary basis."

Where, then, does Thailand fit into the global struggle for gender equality? For centuries, cultural customs in Thailand have skewed toward strictly defined gender roles, often inhibiting the advancement of women. The old Thai proverb "the husband is the forelegs of the elephant and the wife the hind legs" implies that men are leaders and women followers. However, Thai women have made enormous strides in the past few decades. Perhaps no other figure illustrates this as well as Thailand's ranking in the world's top ten

THAILAND'S GENDER INEQUALITY RANKING

The Gender Inequality Index (GII) is an essential part of the United Nations Development Programme's yearly Human Development Report. The latest report, published in 2015, ranks gender inequality in 155 nations for the year 2014. Rankings are based on three aspects of human development: reproductive health (measured by maternal mortality and teenage birth rates), empowerment (measured by the proportion of parliamentary seats occupied by women and the prevalence of secondary education in women over the age of 25), and economic status (measured by the proportion of women 15 years of age or older who are actively employed). A ranking of 1 indicates the country with the least gender inequality in the world.

Thailand's ranking has remained fairly constant in the GII rankings since the first report was created in 2010, hovering roughly around "better than half the world." However, since 2011, Thailand has shown improvements in a decreasing teenage birth rate and increasing numbers of women involved in secondary education.

Here are the available 2015 GII rankings of ASEAN nations:

Country	Ranking	Country	Ranking
SINGAPORE	13	PHILIPPINES	89
MALAYSIA	42	MYANMAR	85
VIETNAM	60	CAMBODIA	104
THAILAND	76	INDONESIA	110

Sources: Human Development Report, UNDP, 2015

(No data available for Laos or Brunei)

ABUSE OF WOMEN IN THAILAND
(Statistics for year 2013)

31,866	Women and girls physically and sexually assaulted
87	Number of females assaulted every day
186	Deaths due to domestic violence
4,000	Rape cases reported to police
2,400	Arrests made by police

Sources: Thailand Institute of Justice (TIJ) and UN Entity for Gender Equality and the Empowerment of Women (UNWOMEN), 2013

countries with the highest number of female executives. Few other nations can also boast such a high workforce participation rate for women: 64.3 percent of females aged 15 and above are employed.

In other crucial respects, Thai females enjoy quite a few essential freedoms and protections that women in many other developing countries can only dream of. Married women are free to adopt titles or family names, according to their preference. The rights of women to file for divorce and to gain custody of their children and assets have also been recognized in Thailand.

However, without adequate access to decent education, these other gains would not add up to much. Thanks to the national policy on eradicating gender disparity at all levels of education, both girls and boys are entitled to 15 years of free education. In areas like increasing literacy these policies have earned top marks. In 1994, nearly two-thirds of the illiterate population were women. Today, among 15 to 24 year olds, the rate of literacy is almost identical for both genders at around 98 percent, a significant achievement.

Over the years, the state has identified other benchmarks of gender equality. As part of the 10th National Economic and Social Development Plan of 2007 to 2011, the Women's Development Plan cemented five key pillars to promote the advancement of women and gender equality: mobilizing all stakeholders to advocate gender equality; enhancing female participation in the policymaking process; improving healthcare services; strengthening women's rights to human security; and fostering more economic participation among women.

However, for all these policies and successes, the gender gap remains as deep-seated as any cultural prejudice. Still seen as caregivers in many ways, Thai women often bear a disproportionate amount of the household chores, from cooking and cleaning to childrearing. Many shoulder these burdens while working full-time. Thai women, like those in other Asian countries, still earn less than men in every major sector. Women are also far more likely to be employed part-time or on a temporary basis.

Nowhere is the disparity between the sexes revealed in more shocking fashion than in the statistics of sexual abuse and domestic violence. A 2005 World Health Organization (WHO) study found that 41 percent of women in Bangkok who had ever been married, lived with a man or had a regular sexual partner had experienced physical or sexual violence at the hands of their intimate partner. Of the women who had been physically abused, 37 percent had never spoken to anyone of the abuse, and only 20 percent had sought help from official channels, such as the police or health workers. Roughly a decade later, in 2013, the Ministry of Public Health's One Stop Crisis Center Report revealed that almost 32,000 women and girls were battered and/or sexually assaulted that year, amounting to 87 cases per day. As a result, Thailand was ranked 36th among 75 countries in acts of physical violence committed against females,

> "Women and girls are not passive and helpless beneficiaries. They are first and foremost solution makers, an army of peacemakers and game changers the world is yet to fully engage."
>
> -UN Women Executive Director Phumzile Mlambo-Ngcuka at the Annual Session of the Executive Board, June 27, 2016, New York

and seventh out of 71 countries for sexual assault, according to the Thailand Institute of Justice. And yet, while the problem is so prevalent and widespread, enforcement of domestic violence laws remains inconsistent and, according to some researchers, biased against women.

In the media too, violence against women has been normalized – the Ministry of Public Health released data in 2015 that revealed a majority of Thai television dramas featured scenes depicting sexual assault. According Jasmine Chia's article in the *Harvard International Review*, this is just one part of the socioeconomic and political machine that portrays Thai women as sexual objects as a matter of course, and which has historical roots in the sex industry.

Chia argues that prostitution in the kingdom was largely a domestic creation. During Thailand's period of rapid modernization, many women found themselves displaced from traditional agricultural life and under pressure to earn more for their families while being shut out of industrial jobs. As a result they turned to sex work as a reliable form of income. Later, when prostitution was criminalized in 1996, it was only the sex workers who were penalized, not the (largely male) clientele.

In this way, the government was able to appear to uphold conservative Thai values without curtailing a booming industry that contributes an estimated US$6.4 billion in revenue a year. As 70 percent of sex workers'

Landmark Legal Moments for Gender Equality in Thailand

› The 1974 constitution was the first to mention gender equality (Section 28).

› Thailand ratified the Convention on the Elimination of All Forms of Discrimination Against Women (CEDAW) in 1985 and its Optional Protocol in 2000.

› Thailand's maternity leave law, implemented in 1993 as part of the Labor Act, allows 90 days of leave. For the first half, mothers receive full salaries. If they choose to take an additional 45 days of leave, they can do so at 50 percent pay. Those working in the informal sector receive no such benefits.

› Thailand endorsed the Beijing Platform for Action in 1995 during the Fourth World Conference on Women. The declaration stated 12 key areas where urgent action was needed to ensure gender equality.

› In 2000, Thailand embraced the Millennium Development Goals (MDGs). Significant efforts have been made to integrate these international principles and instruments into policy and programing frameworks, as evidenced by the Constitution B.E. 2550 (AD 2007), which contains provisions for anti-sex discrimination and gender equality. The Domestic Violence Victim Protection Act B.E. 2550 was also established.

› The Gender Equality Act came into effect in September 2015. The law aims to eliminate discrimination among the sexes and is the first Thai law to contain language explicitly recognizing gender diversity. Additionally, a special committee has been set up to promote parity and mediate on cases of gender discrimination among the sexes, including "any act or failure to act which segregates, obstructs or limit any rights, whether directly or indirectly, without legitimacy because that person is male or is female or has a sexual expression different from that person's original sex." Offenders could face as much as a six-month jail term or a fine of up to 20,000 baht.

Gender Equality Means LGBTI Too

Gender equality is not just about male-female heterosexual relations. The wide umbrella of "gender issues" also includes the rights and equality of the LGBTI community (lesbian, gay, bisexual, transgendered and intersex persons). Thai society is famous for its acceptance of homosexuality and gender diversity including *kathoey* ("transgendered"). According to the Ministry of Pubic Health, Thailand is home to an estimated 600,000 gay men. The country has also become a dream destination for gender reassignment surgery. This degree of Thai tolerance, inspired in part by Buddhist compassion, has also lent a veneer of acceptance to those who are LGBTI. But for all the superficial tolerance, members of the LGBTI community still face discrimination in the workplace and are frequently depicted through negative stereotypes in the media. However, being accepted by their own flesh and blood remains the biggest struggle they face in this traditional, family-centric nation.

The protection of women's rights and those of the LGBTI community are deeply intertwined. For example, Thailand's Gender Equality Act, which came into effect in September 2015, not only guarantees women legal protections but is also the first Thai law to explicitly recognize gender diversity. As Thailand's first legislation that protects LGBTI people from discrimination, it is a landmark document for gender equality.

With more female and LGBTI representation in the corridors of political power, the massive gains that Thai women have made in fields like education and economics could be further bolstered to bring about a much fairer society for all people, no matter where they fall on the spectrum of sexuality or gender identity.

---

clients are Thai men (not foreigners as is sometimes suggested), this gave rise to a generation of men who were socialized under policies that demonized the "corrupting" influences of sex workers yet validated consumerism of the sex trade. This has only fed into the normalization of women as sexualized objects in Thai society — and often contributed toward violent objectification.

However, Thai government bodies, private foundations and community organizations have begun to implement programs to empower women, address discriminatory policies and foster self-reliance, especially in marginalized communities.

With **gender mainstreaming** also becoming the norm among public policymakers, the tides are turning toward a public mindset that appreciates rather than ignores the differences between the sexes, bringing them closer together in a marriage of equals.

STEPS TOWARD A MORE EMPOWERED SISTERHOOD

In 2014, Thailand ranked 76th out of 155 countries in the United Nations Gender Inequality Index, indicating that Thailand enjoys less gender inequality than roughly half the ranked countries of the world. However, women's groups and NGOs maintain that Thailand's growing wealth and vastly improved Human Development Index ranking masks concerns and abuses still plaguing women today, mainly in the realms of domestic violence, sex segregation in work and lack of participation in decision-making.

Gender inequality is often subtle and symptomatic of other related issues, such as poverty, poor education, high divorce rates and lack of resources. In particular, women in marginalized communities such as rural areas, hill tribe regions and the conflict-ridden South, suffer the least access to health, education

---

Gender mainstreaming

A public policy approach that attempts to promote gender equality by evaluating the impacts that proposed legislation and programs will have on both men and women.

and economic resources and are often reliant on their male partners for income generation and decisions about household matters.

In rural areas where agriculture is the main livelihood, women often don't have the kind of employment opportunities that would allow them to contribute more income to their household, or their families become broken when their partner goes to an urban area to seek better work. Sometimes, their partners never return, leaving the women penniless and skill-less with children to support.

Women in these communities can be empowered to take an equal share in economic opportunity by giving them fair access to the resources needed for entrepreneurship and the running of SMEs. Although Thailand enjoys high scores in female participation in the work force, statistics show that women workers still tend to be concentrated in low-paid, low-skilled occupations, especially in rural areas. Together with the private sector and financial institutions, the government can directly empower women to find sustainable employment by offering vocational training, providing access to funding for startups, and supporting the distribution of SME products to the domestic and international market.

There have already been efforts in this field with the creation of such organizations as the SUPPORT Foundation, which was established by Queen Sirikit. The foundation aims to empower women in rural communities to become more self-reliant. The main participants are often single mothers and female heads of households, who are provided with the resources, training and know-how needed so they can earn more income, be more independent, and work with dignity.

Several other successful initiatives offer support to marginalized and conflict-ridden communities in Thailand, though many of these tend to be NGO-run or community-based.

Women take part in a capacity building exercise during a rural self-help group meeting.

Since the early 2000s, for example, Oxfam has joined hands with Unilever and the Deep South Coordination Centre to promote women's leadership in the provinces or Yala, Narathiwat and Pattani. As a result of years of ethnic sectarian violence, more than 3,000 households have lost at least one male member, leading remaining family members to suffer economic hardship. This has forced women to take a leading role in generating income in a society that has traditionally been characterized by gender inequality. The project aimed to empower 9,000 affected women by 2016 to become entrepreneurs. Capital and business capacity building initiatives carried out by Oxfam have launched and scaled up more than 15 female "micro enterprises" that sell high-quality products, including organic rice, preserved seafood, snacks and handicrafts. These businesses have contributed income stability and self-esteem to affected families.

Another successful initiative, The Kamlangjai Project, was founded by Princess Bajrakitiyabha in 2006 to support female inmates in Thai prisons. In part, the project came about to redress the fact that correctional and rehabilitation systems are designed for men, and force women to undergo an

"Gender inequality is often subtle and symptomatic of other related issues, such as poverty, poor education, high divorce rates and lack of resources."

Thailand's thriving sex industry captures many of the contradictions of gender equality.

identical incarceration experience as men. Yet the general profile of women prisoners is often a less violent one – frequently they are convicted for drug-related crimes or theft – and they may have intricate histories of abuse, victimization, trauma and addiction that are different from male prisoners. To address some of the gaps that have arisen from male-focused prison systems, The Kamlangjai Project provides healthcare for women, with an emphasis on assisting pregnant and nursing inmates as well as the children who live with their mothers in prisons. The project also seeks to empower women through vocational training initiatives to minimize their chance of re-offending after release and to provide a buffer against the "double jeopardy" of serving time and being discriminated against.

Based on the success of The Kamlangjai Project, the princess later established what came to be known as the Bangkok Rules, a program built on 70 criteria designed to address the specific needs of women in criminal justice systems. In 2010, the Bangkok Rules were adopted by the UN General Assembly as the United Nations Rules for the Treatment of Women Prisoners and Non-custodial Measures for Women Offenders. Since then, numerous countries and international organizations have put the rules into practice.

In a sense, these programs highlight just how great the need is for other such initiatives focused on women's empowerment, vocational training for women at risk or living in institutions, and training for women in technical and business management skills.

The Department of Women's Affairs and Family Development (DWF) under the Ministry of Social Development and Human Security, for example, has successfully established eight Women and Family Development Centers in vulnerable regions throughout the country to provide support networks and employment opportunities to at-risk women. Short-term training, certificate programs and advanced skills are offered in vocations such as hairdressing, dressmaking, nursing, handicraft production and traditional Thai massage. In a larger sense, however, economic inequality remains a national issue affecting even women in high-skilled jobs. A larger discussion on how to eliminate unequal pay must be followed by strategies and implementation.

In the case of violence against women and girls, too, communities and local stakeholders can affect great change. Given the failed enforcement of national laws against abuse and violence, establishing community-based projects to create surveillance units can help raise awareness and curb abuses at the local level. Already, programs have been implemented to raise public awareness about the prevalence of domestic abuse in Thailand and to encourage individuals at the grassroots level to help monitor cases of abuse and domestic violence. In 2014, the Foundation for Women, Dentsu Plus and Major Cineplex partnered to create a sound check featuring the cries of domestic violence victims to run in theaters before every movie. In this deeply

Thailand's Maternity Leave Policy Allows FOR 90 DAYS OFF

"SUPPORT" for Women

Gender inequality heavily impacts the nation's most vulnerable communities – such as rural people, impoverished neighborhoods and ethnic or religious minorities– because they often don't enjoy the same access to healthcare, education, technologies, social networks and other resources that the wealthy, the privileged and the urban do. This fact has not escaped the notice of the royal family, who has established several foundations and organizations to include marginalized communities in the national goal toward sustainable development.

One long-standing example is the "Foundation for the Promotion of Supplementary Occupations and Related Techniques of Her Majesty Queen Sirikit of Thailand", also known by its shortened form, the SUPPORT Foundation.

Established in 1976 by Queen Sirikit, the foundation is guided by the principle of fostering greater self-reliance and provides supplemental employment to mostly low-income rural women and teenagers by training them in the production of traditional Thai arts and crafts. These products are in turn sold in specialized SUPPORT Foundation stores throughout the country, linking the marginalized or remote groups to a direct market. The queen also promotes their work and their handicrafts by wearing them and bringing them with her on trips abroad. While this has provided an additional income stream to indigent families (many of them headed by women who, without this opportunity, may have taken up riskier employment such as migration to urban areas or joining the sex trade), the program has also had the added benefit of preserving a rich national heritage of arts, crafts and traditional vocations that are important to a Thai sense of identity.

Many of the trainees participating in SUPPORT are, in fact, single mothers, female heads of households and teenagers. Through the program, these women are not only provided with tangible, income-earning skills in such folk crafts as silk weaving, silver and gold smithing and basketry, but they also gain a sense of pride, empowerment and self-reliance. With these new skills, marginalized women are better equipped to determine their own future. Those who produce exceptionally fine work also have the opportunity to become SUPPORT instructors.

The royal family has started numerous such projects aimed at poverty alleviation, women's empowerment and social justice, not least of which are Princess Maha Chakri Sirindhorn's Sai Jai Thai Foundation and Phufa stores. While the Sai Jai Thai Foundation targets military personnel, police officers and civilians who were hurt in the line of duty, the Phufa stores aim to raise rural people out of poverty. Although not specifically focused on gender equality, both of these projects address issues of social justice, which feeds into the advancement of all marginalized groups – including women.

Established in 1975, Sai Jai Thai offers both a stipend and vocational training in leather and glass artisanship to the handicapped to compensate veterans for their

Queen Sirikit (right) founded the SUPPORT Foundation in 1976 to help empower women and girls in rural communities across Thailand.

sacrifice and to encourage sufficiency living. Sai Jai Thai crafts are sold in gift shops in high-traffic locations, such as Suvarnabhumi Airport in Bangkok.

On the other hand, the Phufa stores are high-quality handicraft shops that carry region-specific goods sourced from rural and impoverished communities. Similar to the queen's SUPPORT Foundation, the Phufa stores offer marginalized communities a distribution channel and encourage the use of local resources and fair trade to give vulnerable communities employment opportunities and greater self-reliance.

With such examples set by the royal family time and again, surely Thailand's governing bodies can also conceive of its own ways to level the social playing field, address economic inequality, address gender issues and hand back more power to the people. •

The Rise of Female Executives

Few figures illustrate the financial betterment of women's lives in Thailand more positively than the huge number of female executives and entrepreneurs. In 2012, Thailand topped the global list of women running their own companies, according to the US-based Global Entrepreneurship Monitor (GEM) survey, which also said that 12 women started or ran their own businesses for every ten men. Although the rate has dropped a little in recent years, it remains remarkably high, with Thai women accounting for about one-third of the board members of companies registered with the Ministry of Commerce.

To commemorate International Women's Day on March 8 every year, Grant Thornton, a London-based professional service network, releases its annual *International Business Report*, which includes a survey of women occupying executive roles. In 2015, Thailand came in fifth with 37 percent, well above the global average of 24 percent, and only a little behind ASEAN leader the Philippines (39 percent).

Sumalee Chokdeeanant, an assurance partner at Grant Thornton Thailand, said, "In Thailand and in many places in Southeast Asia, it is not unusual for women to be in senior executive roles. Asia is a strongly family-orientated society and the female has always played a key role in the household, especially governing finances. Over time, as business has grown, this has simply become the norm for us in Thailand. This is evidenced by a continuing growth trend of women in executive roles here." She added that the integration of the ASEAN Economic Community means there will be more golden opportunities for women to take on senior managerial roles in the years to come.

In the domestic labor market, female participation has been on the rise for many years, thanks in no small part to the enactment of the Labor Protection Law of 1998. That law stipulated gender equality in employment, health security, work safety and the prohibition of sexual harassment in the workplace. This legislation, in tandem with a raft of governmental measures to boost the education levels of girls and their access to tertiary institutes, has resulted in what the UN's Gender Equality Index has called one of the highest "labor force participation rates" for women in the world: a little over 64 percent. Thai females also account for about one-third of the board members of companies registered with the Ministry of Commerce.

For Thailand and other nations, the more balanced the workplace and the executive boards are between men and women, the more well informed are the decisions that form the axis upon which these businesses pivot. Francesca Lagerberg, global leader for tax services at Grant Thornton, said: "That greater diversity in decision-making produces better outcomes is no longer up for debate. For businesses, better decisions mean stronger growth, so it is in their interests to facilitate the path of women from the classroom to the boardroom."

moving way, audiences were made aware that domestic violence is a social issue, not a family one, and were encouraged to report sounds or signs of domestic violence to a hotline. Other awareness-raising and self-monitoring programs like this have cropped up on the community level in Thailand, spearheaded by local and national NGOs.

Public participation of women is also often easiest achieved initially at the local level. If women are encouraged to take leading roles as advocates, organizers and educators among local communities, gendered issues that are often kept hushed up due to social stigmas — such as domestic violence, trafficking and forced labor — will have the chance to take their place on the political agenda. This opens the door for gender issues to become priorities on provincial and, ultimately, national levels.

As the primary caretakers of families, improving reproductive health and family unity will also greatly impact women's quality of life

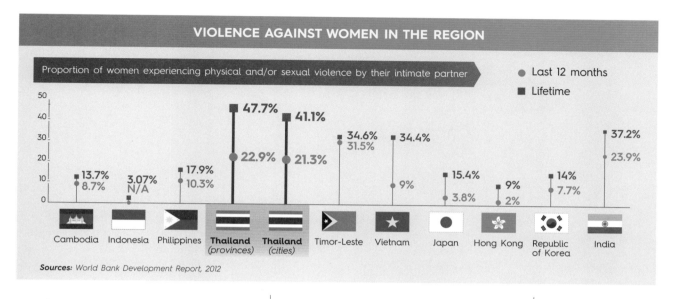

as well as the health of society at large. With the divorce rate on the rise, single motherhood is an increasingly common status in Thailand and is becoming a crucial issue for sustainable development. However, divorce settlements often favor the father and leave single mothers with little economic support. Thus bereft, many single mothers live in poverty, unable to secure the educational, health and economic resources that guarantee healthy child development. This, in turn, inhibits robust sustainable development on a national level. Despite all this, Thailand currently has no direct welfare system that benefits single mothers.

Although the Ministry of Social Development and Human Security has launched a number of campaigns to discourage teenage pregnancies and support single mothers, more can be done within the community. The DWF has established more than 7,000 family development centers throughout the nation with the primary aim of strengthening family unity. The development centers survey families to identify problems and types of need – such as single-parent households, elderly citizens living on their own, insufficient income, drug abuse or domestic violence – and offer advice, activities or monitoring to support and strengthen the household. This is an important national priority, as the government considers families the basic units of society in which all development – whether social or economic – begin. In addition, the centers address gender inequality specifically by rooting out discriminatory practices at home, encouraging family members to shoulder burdens and share benefits equally.

As much as gender equality is and should be an important agenda for national policymakers, it is also a deeply personal matter that permeates the intimate fabric of family life, personal health and community relations. While governmental bodies should set the tone for gender equality, the main source of the problem – which lies in cultural prejudices – must be changed in the minds and attitudes of individuals, one at a time.

Further Reading

• *Gender Mainstreaming in Environment and Sustainable Development Projects: A Perspective from the Asia-Pacific Region* by UNDP, 2015

• *Driving the Gender-Responsive Implementation of the 2030 Agenda for Sustainable Development* by UN Women, 2016

• *Gender Equality and Human Rights* by Sandra Fredman and Beth Goldblatt, UN Women, 2015

6 CLEAN WATER AND SANITATION

Ensure availability and sustainable management of water and sanitation for all.

CLEAN WATER AND SANITATION
Managing a Fundamental Element

Calls to Action

- Find sustainable ways to meet increasing demand for water as a result of population growth and economic expansion

- Build upon the work of King Bhumibol Adulyadej to promote integrated water resource management among responsible agencies and improve mechanisms to mitigate the effects of droughts, floods and irregular rainfall patterns

- Strengthen the participation of local communities in improving water resource management and sanitation

- Increase capacity to treat wastewater and crack down on companies dumping untreated waste into the water supply

- Educate individuals to use water in a more conservative manner and avoid practices that pollute water resources

- Combat threats that are damaging key watersheds

Completed in 2005, Lat Pho canal (right) in Samut Prakan has helped mitigate the impacts of seasonal flooding.

Across the globe, around four billion people face severe water shortages for at least part of the year. Glaciers are continuing to shrink, aquifers and lakes are drying up and less water is flowing in our rivers. Quite simply, our planet's water supply is disappearing at an alarming rate. As a result, in the not-too-distant future, we may see severe water scarcity leading to more frequent crop failures and low production yields, which may cause food price increases, famine, widespread starvation and serious conflicts between nations and peoples.

Thailand is fortunate to have been endowed with an abundance of water resources due to its geographical position in the tropics, which feature monsoons that result in a six-month long wet season. According to "Droughts in Thailand", a 2012 report prepared by leading Thai experts, the average precipitation in Thailand is 1,374 millimeters per year, well above the global average of 990 millimeters. And overall, Thailand fares well on Goal 6, according to the indicators used by the SDG Index to measure its performance. Access to improved water is enjoyed by nearly 98 percent of the population and 93 percent have access to improved sanitation. Thailand also already meets the threshold for freshwater withdrawal. But supply is not the issue. Instead, resource management remains the key challenge facing the kingdom.

To help manage the ebb and flow of water, adequate irrigation infrastructure – such as dams, dykes, moats, canals and floodgates – covers only a small fraction of the country's land area, and the distribution of these facilities is uneven. Overall, only one-sixth of the country's farmland is irrigated, according to the Hydro and Agro Informatics Institute (HAII) under the Ministry of Science and Technology. The vast majority of that lush farmland is in the central region, famous as the country's "rice bowl". Meanwhile, farmers in other parts of the country – where rainfall can be unreliable, waterways non-existent and irrigation systems rare – face the regular threat of drought and must rely on whatever strategies they can devise to eke out a living.

In some cases, projects carried out by the Royal Irrigation Department, HAII, or various organizations such as the Utokapat Foundation have helped to improve access to water. Grassroots water resource management has also taken off across the kingdom, with small communities banding together to carry out measures that ensure they have water reserves during dry spells.

As an example, Ban Limthong community in Nangrong district of Buriram province had for years faced recurrent water shortages. To address their water issues, local villagers devised a community level water resource management plan taking into account the area's unique needs. As a part of this, they established a community water committee, dug canals and retention ponds, built irrigation and water treatment systems, and set up water user accounts.

For decades, successive governments have also advised farmers in these drought prone areas to diversify away from water-intensive crops, the most demanding example being rice. In 2016, these efforts were ramped up and state-funded training sessions carried out around the country to educate farmers on alternative crops like fruit trees and peas. Drought, however, is only one part of the triple threat that beleaguers the country's authorities on water management.

Dealing with floods and treating wastewater are the two other key issues. The lower part of the central region, irrigated by a low-lying river basin and floodplain, is especially susceptible to rising waters and is often inundated by run-off from northern waterways. During the rainy season, these rivers often swell, bursting their banks and overflowing into urban areas.

In 2003, the Royal Irrigation Department embarked on a mission to transform the small Lat Pho canal in Samut Prakan into a viable tool to mitigate the impacts of seasonal flooding. Initially, the canal was only 10–15 meters wide at a depth of just 1–2 meters. Following advice from King Bhumibol Adulyadej, it was

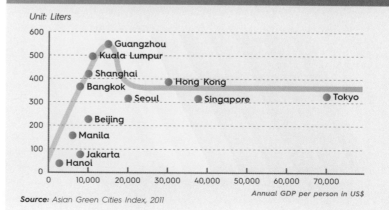

WATER CONSUMPTION IN LITERS PER PERSON PER DAY

Source: Asian Green Cities Index, 2011

The King and the Water Paradox

"**From that point on,**" King Bhumibol Adulyadej wrote in his journal, "I have thought about this seemingly insoluble and paradoxical problem: when there is water, there is too much, it floods the area; when the water recedes, it is drought."

"That point" the king referred to was a 1955 tour of Thailand's Northeastern region, the country's poorest, where farmers were barely able to produce one successful crop. Traveling over 680 kilometers by train and 1,592 kilometers by road, flooded by stories and images of hardship, the king was moved. The effective management of the country's water resources has been an abiding preoccupation for him ever since.

Over the ensuing decades, a wide variety of resources, ideas, projects, programs and technologies were directed toward improving this fundamental underpinning of rural existence. "If we take proper care of the environment, there will be water for many hundred years", he once observed, echoing the long-term thinking that is at the heart of sustainable development. "By that time our descendants might be thinking of some new ways to solve the problems." As much as any story of his reign – the rise and fall of Communism or the rising skyline of Bangkok – the king's attempts to improve the security of the farmer was the one he personally led.

In 1962, near his summer residence of Hua Hin, King Bhumibol invested his own money to build an earth-fill dam near the coastal village of Khao Tao. The poor living conditions of the villagers were due, in part, to soil erosion and fresh water scarcity. The dam blocked seawater from flowing into a natural channel and helped trap rainwater flowing in. The 600,000 cubic-meter-capacity reservoir known as Khao Tao reservoir would be the first of many localized projects the king helped create. It was also typical of his development model: establish direct contact with the villagers and learn about their problems, offer practical solutions the villagers themselves could adopt, cut through bureaucratic red tape to implement a solution, and then monitor the progress.

In order to help villagers better manage their natural water supplies, throughout the country he advised them to protect watersheds and build weirs as miniature check dams to help regulate a river's flow. He also advocated the construction of reservoirs and ponds dubbed "monkey cheeks" to store floodwaters during the rainy season so

King Bhumibol Adulyadej spent decades addressing water management.

they could be used later when other sources had dried up. The idea, which came to him in the wake of the 1995 floods, was inspired by a childhood memory and based on the way that monkeys keep special reserves of food in their cheeks to use in times of emergency.

"A monkey can keep almost an entire bunch of bananas inside its cheeks," he recalled at the time. "Only later will the monkey take the bananas out, chew and swallow them." In essence, the ponds that the villagers dug to capture floodwaters functioned in the same way.

Known as *kaem ling* in Thai, the concept has gained notoriety nationwide as a cheap and environmentally friendly way to manage water and mitigate the effects of floods and drought. Today there are some 190 monkey cheek water retention areas across Thailand and in the wake of the 2016 drought the government has called for the construction of more *kaem ling* projects.

When coming up with such plans the king's modus operandi has been to deploy "appropriate technology" that harnesses the raw power of nature. By the 1990s, with help from the king's Chaipattana Foundation, this recipe had been repeated successfully thousands of times. •

widened to about 65 meters and deepened to seven meters, meaning that water flowing from the Chao Phraya River now only has to travel 600 meters (instead of 18 kilometers on its natural course) to reach the Gulf of Thailand. In addition, four floodgates were installed to mediate the flow of drainage. The net result is that by draining off around 40 million cubic meters of water a day, authorities are able to maintain lower water levels in the lower Chao Phraya Basin. The floodgates can also be closed during drought or high tide to prevent the influx of seawater into the Chao Phraya River.

Thailand's last major flood was in 2011 when unusually heavy rainfall – 20 percent more than the average – combined with mismanagement and perennial problems, such as flagrant land misuse and outdated drainage systems, to set the stage for what the World Bank called the fourth costliest disaster in history. As is often the case, such disasters can serve as a catalyst for positive change.

In 2012, Thailand tapped Japan for assistance in developing flood warning systems. The Project for the Comprehensive Flood Management Plan for the Chao Phraya River Basin, carried out with the assistance of the Japan International Cooperation Agency, led to the creation of a unique flood inundation forecasting system that can provide the public with real-time flood warnings up to seven days in advance. Additionally, in May 2016 the National Disaster Warning Center announced that it had conducted a successful trial of a system developed by NEC Corporation, which had effectively predicted flooding in Uttaradit province during a period from November 2015 to March 2016. In the case of impending floods, such alerts make it possible for companies to move equipment and product to safer locations. It allows farmers to get livestock to higher ground or harvest crops before fields are inundated, and it gives residents ample time to prepare sandbags and other countermeasures to protect their homes.

Yet, another looming threat in the form of climate change holds the potential to deliver far more dire water-related disasters in the future. Among climatologists, the consensus is that water will be impacted more than any other element by global warming. Rising sea levels have already begun inundating Thai coastal areas with saltwater, contaminating both farmland and freshwater sources, according to Dr Royol Chitradon, director of the HAII, which monitors data on water and advises other agencies on management issues. The most worrisome aspects of climate change are its volatility and unpredictability. Entire weather patterns are changing, and the cycle of floods and droughts may become even more severe and erratic in the future. Such changes, said Dr Royol, will make it harder for the powers-that-be to predict and map out accurate water management plans. "The problem is that rain will fall in unusual places at unusual times. We might not be able to use old models and analytical methods to predict these kinds of patterns in the future", he said.

> *Water Will Be Affected* MORE THAN ANY OTHER ELEMENT BY GLOBAL WARMING

Massive floods inundated much of Thailand in 2011.

As successive governments struggle to come up with a comprehensive water management plan to forecast such dilemmas and ward off future floods and droughts, the subject of water management in Thailand has taken on a grave and tremendous significance: part science, part prophecy and all important. One thing is certain: firm policies and plans are needed while there is still time. For starters, government agencies and other relevant parties need to work in a more integrated manner to address the issues at hand. The emphasis during both the short- and long-term should be on implementing systems to mitigate the most severe effects of droughts, floods and climate change – and to make sure individuals have the information necessary to do their part.

WASTEWATER WOES

As Thailand continues to industrialize and urbanize, it must be more proactive in dealing with the rising tide of untreated wastewater. According to the Asian Green City Index of 2011, Bangkokians use 340 liters of water per person per day, higher than the 276-liter average of residents in the 22 other Asian cities surveyed. Because of a lack of wastewater treatment plants, the discharge (mostly untreated) goes straight back into the Chao Phraya River and various canals.

On a visit to Thailand in 2013, the UN Special Rapporteur on the human right to safe drinking water and sanitation Catarina de Albuquerque expressed shock over the amount of "untreated human waste" being discharged into the country's rivers and other water sources. Only about 20 percent of the wastewater produced daily in the kingdom is treated. In Bangkok, about 40–50 percent is treated but outside the capital the share is much lower. This statistic is significant as it also highlights the issue of unequal access to clean water.

While less than one percent of Thailand's population had access to water and sanitation in 1960, the country reached almost universal coverage by 1999. However, there are strong disparities in quality and the "majority of Thai people are not provided with safe drinking water", according to the UN Special Rapporteur. Indeed, according to estimates by the Ministry of Public Health, the current rate of access to safe drinking water in rural and urban areas is only about 25 and 40 percent respectively. In an effort to level the field, the government aims to provide more than 7,000 villages throughout the country with access to clean water by 2017.

In addition to untreated fecal matter, wastewater discharge from industrial, mining and agricultural projects is exacerbating water contamination. By law, factories are required to install their own treatment systems but such rules are rarely enforced, so many factories simply discharge waste directly into water sources. As a direct consequence, the water quality in many of Thailand's largest rivers and canals has dipped considerably. Bangkok's canals, which feed directly into the Chao Phraya

> "Government officials have admitted that nearly half of the country's wastewater treatment facilities are offline or experiencing technical difficulties, which is why at present capacity only a fifth of wastewater is being treated."

Volunteer prison workers remove sewage and plastic waste from drains outside Bangkok.

The Imbalances of Thai Holidays

Revellers throw water during Songkran in Bangkok.

For Thais, water is much more than just a natural resource. It's integral to the lifeblood of Thai culture, coursing through the Songkran celebrations in April when Buddha images are washed and revellers doused, at the Loy Krathong festival where offerings are made to the Goddess of Water, and in rituals like the anointing of a bride's and groom's hands. The irony is that while water is considered sacred, some Thai holidays like Songkran and Loy Krathong have evolved to include environmentally unfriendly practices that result in negative impacts on the nation's water resources. In that sense, they have become a paradox.

Songkran, Thailand's nationwide New Year's water fight, has become famous around the world. But with its popularization has come an influx of tourists eager to take part and a growing number of Thais who also see "playing water" as an opportunity to let loose. While it's certainly fun, the underlying problem is all the water that ends up wasted.

In 2016, as the country experienced its worst drought in decades, the government experimented with enforcing a 9pm curfew and limiting the Songkran revelry in the capital to just three days. By doing this in Bangkok alone, the government estimated it could prevent the loss of as much as five billion liters of water.

Now, it's hard to estimate exactly how much fresh water is lost each year during Songkran and it's even harder to picture what five billion liters of water would actually look like. So for reference, consider this: a square block of water measuring 10 x 10 x 10 meters would contain one million liters. That means at 10 meters deep and 10 meters wide, the 5,000 such blocks required to contain five billion liters of water would stand 50 kilometers high. And remember, that's just the amount the government estimated it could save in Bangkok.

In addition, instead of preserving Thai culture and promoting a real sense of community, Songkran is now a money-maker, branded as a party and fueled by indulgent drinking. Road fatalities, one of the areas Thailand scores very poorly on in the SDG Index, notoriously spike during this time. In 2016, 442 people died on the roads during a seven-day period, the worst result in the previous ten years, showing how the imbalances featured in modern-day Songkran celebrations have serious consequences.

Loy Krathong, which Thai Buddhists celebrate by placing flowers, candles and joss sticks into a small vessel and floating it down a river to pay homage to Buddha and Ganga, has also seen changes in practices over the years. Traditionally, the vessels were made from banana trees and decorated with flowers, but in recent years Bangkok city authorities have found that a substantial number of *krathong* vessels are made from plastic or Styrofoam.

In 2015, the number of plastic or synthetic foam *krathongs* placed into the Chao Phraya River alone was estimated to be more than 71,000 out of about 825,000 vessels. Many of the vessels are collected not long after they are released, but a significant number end up contributing to the refuse polluting the country's waterways.

While the commodification of Thai holidays may seem like something an individual has little control over, Songkran and Loy Krathong revelers do ultimately make their own decisions about how to celebrate these holidays. If each person tried to demonstrate more balance in conserving water resources, respect the environment and think more sustainably during these holidays, it would go a long way toward reducing negative impacts.

For example, instead of using synthetic materials to make *krathongs*, environmentalists and the Natural Resources and Environment Ministry are encouraging people to use biodegradable materials during future Loy Krathong festivals. Some alternative construction materials include bitter gourd, papaya, pumpkin, banana trees and, of course, real flowers. Bread is also a good alternative as it can be gobbled up by fish.

River, in particular, are cause for serious concern and would benefit from more regular dredging. In particular, plastic waste has become a major scourge for Bangkok's network of pumping stations, clogging machinery and hindering drainage during seasonal downpours.

One of the main obstacles to improving water quality is that the kingdom has only about 100 wastewater treatment plants. Unfortunately, not all of them work properly, or at all. Government officials have admitted that nearly half of the country's wastewater treatment facilities are offline or experiencing technical difficulties, which is why at present capacity only a fifth of wastewater is being treated.

The good news is that Thailand is in the midst of constructing several new treatment plants. The former director of the Pollution Control Department, Supat Wangwongwatana, whose recommendations to other ministries paved the way for new laws cutting down on motor vehicle emissions, says that money for these wastewater treatment plants must come from government coffers. But maintaining them should be the responsibility of the municipalities who must collect fees paid by local communities or businesses, under the "polluter pays" principle of international environmental law.

This approach of providing economic incentives or disincentives, as the case may be, is the only way to get the public interested in recycling and reducing their own waste generation and overall water usage. Appealing to the bottom line is a strategy that can also pay dividends when dealing with companies and plants that illegally dispose of hazardous wastewater (a wastewater fee already applies to hospitals, hotels and businesses, at between US$0.13 to US$0.16 per cubic meter). However, there are major gaps in monitoring and until these types of programs and policies are implemented and enforced, water contamination will continue to be a swelling issue.

CUTTING THROUGH RED TAPE

Given how much time and effort have been put into the issue over the decades, the management of water resources in Thailand's rural areas should not be such a daunting task. But in practical terms, Thailand still struggles in this area, with bureaucratic "red tape" serving as the primary obstacle to making water management more efficient.

Unlike most countries, Thailand does not have a law governing the management of water resources and thus cooperation in this area is highly fragmented between many bodies. Currently, 30 different state organizations under seven different ministries apply more than 50 different laws to the multiple issues of water management. Bureaucratic obstacles, from overlapping authority to disunity within the various departments responsible for water management, end up delaying the decision-making process and impeding budgetary

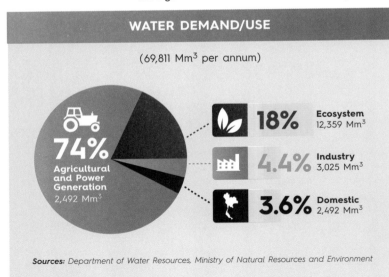

WATER DEMAND/USE

(69,811 Mm³ per annum)

74% Agricultural and Power Generation 2,492 Mm³

18% Ecosystem 12,359 Mm³

4.4% Industry 3,025 Mm³

3.6% Domestic 2,492 Mm³

Sources: Department of Water Resources, Ministry of Natural Resources and Environment

disbursements. It can also result in flabbergasting gaps in execution that sometimes leave beneficiaries high and dry.

For example, in some cases reservoirs are built in villages but without the installation of accompanying pipework to siphon water into agricultural areas (often, this is due either to budgetary constraints or a lack of foresight). Meanwhile, the maintenance of reservoirs and distribution systems is even more complicated due in part to the misuse of budgets. According to the law, local people cannot carry out repair work by themselves because the reservoirs are government-owned, and while some local organizations may have the budgetary resources to carry out repairs, they more often than not lack the technical skills to do so. It should come as no surprise then that thousands of Thailand's reservoirs and irrigation systems in agricultural areas are in a ruinous state.

What's needed is a multi-stakeholder approach that cuts through traditional red tape, while collaborating with communities, local government administration and other relevant actors, such as civil society organizations that are often willing to help. To that end, the Royal Initiative Discovery Foundation embarked on a collaborative quest in 2014 to develop what it dubbed a "Nan Model" to demonstrate how to achieve effective, efficient restoration of reservoirs and associated structures.

In Nan province, where the mountainous terrain renders large-scale irrigation systems almost impossible to install, local people struggled for years to make due with inadequate small-scale reservoirs. The concerned authorities had also never developed appropriate flood management systems, meaning that during droughts, farmers had insufficient water for crops, while almost every rainy season locals faced flash floods that inundated villages. Lack of foresight also led to failures like the Nam Lieb reservoir in Chiang Klang district. While the reservoir was completed in the 1980s, effective irrigation systems were never built to supply water to nearby villages, resulting in chronic water shortages and low agricultural yields.

Then in 2014, the Royal Initiative Discovery Foundation collaborated with the Nan provincial governor's office, local stakeholders and other organizations to repair and service small-scale reservoirs and irrigation systems across the province in an effort to replenish water resources.

As part of this Nan Model initiative, the foundation and relevant authorities conducted repair work on 560 reservoirs and other related structures, some of which had been inoperable for decades. The Nan Model encompassed construction of pipework that enables locals to manage water resources on their own, development of water management plans to boost productivity, and raising awareness about the benefits of reforestation and the adverse effects of slash-and-burn agriculture.

Overall, more than 41,700 households are to benefit from the Nan Model project, with newly restored infrastructure supplying an adequate amount of water to more than 16,000 hectares of cropland.

Bureaucracy remains a huge impediment to improving water resource management. But as the Nan Model demonstrates, through cooperation and collaboration that involves all relevant stakeholders, progress in this key area is entirely possible. To help cut through the red tape that binds up the decision-making process and final implementation, the government needs to streamline and consolidate bureaucratic processes and legal apparatuses, decentralize authority in terms of which bodies have final approval over local projects, and draw up provisions that allow for the fast-tracking of approval and implementation of urgent projects.

The Nan Model Benefits MORE THAN 41,000 HOUSEHOLDS

Further Reading

• *History of Water Resources and Flood Management: A Policy Brief* by Duenden Nikomborirak and Kittipong Ruenthip, 2013

• *Thailand Flooding 2554 Rapid Assessment for Resilient Recovery and Reconstruction Planning* by the Ministry of Finance, Royal Thai Government and the World Bank, 2012

7 AFFORDABLE AND CLEAN ENERGY

Ensure access to affordable, reliable, sustainable and modern energy for all.

AFFORDABLE AND CLEAN ENERGY

Thailand's Opportunity to Create a Low-Carbon Society

Solar panels are used to generate electricity in Prachuap Khiri Khan province on the Gulf of Thailand.

Calls to Action

- Achieve energy self-sufficiency by scaling up and developing alternative energy sources as a means to decrease reliance on imported energy and compensate for a diminishing domestic natural gas supply

- Increase R&D and support innovative new technologies such as electric cars through appropriate infrastructure and funding

- Create an energy-saving mindset across the entire population including adults, students and businesses to help reduce overall energy demand

- Sincerely address grassroots concerns about the negative impacts of extractive projects

Ever since Thailand began industrializing in the 1960s, its steady economic growth has been largely backed by fossil fuels such as coal, oil and natural gas. While the peak levels of growth of the late 1980s and 1990s eventually bottomed out, the country's energy demand has consistently risen.

On the plus side, proactive initiatives launched in the 1980s to aggressively expand the nation's energy infrastructure mean that nearly 100 percent of the population has access to electricity today. However, on the negative side, the resulting spike in demand has increased the kingdom's reliance on fossil fuels, which still account for 85 percent of domestic energy consumption. This fossil fuel dependence is somewhat worrying, and here's why.

Because of the CO_2 emitted by fossil fuels, their widespread global use poses a considerable threat to the planet. For every ton of oil-equivalent energy used, 2.4 tons of CO_2 emissions are released. In the past 150 years, the planet's CO_2 concentration has jumped from 280 ppm (parts per million) to 400 ppm. As The Earth Institute director Jeffrey D. Sachs notes, "We have reached a level of CO_2 in the atmosphere not seen for the past 3 million years!" If that sounds frightening, it should. Climate experts warn that a rising CO_2 concentration directly correlates to a marked rise in global temperatures, and that we could soon be living on a planet that is on average two degrees Celsius warmer than before the Industrial Revolution. Such a drastic shift in the Earth's climate system may change the current patterns of rainfall and lead to heat waves, severe droughts, mega-floods, extreme storms, crop failures, a massive rise in the sea level and acidification of the oceans.

But we have alternatives to fossil fuels — for example, far less CO_2 is released during the generation of solar, wind and hydroelectric power. The technologies and systems to

Feed-in tariff (FIT)
The world's most commonly used measure to promote renewable energy offers a guaranteed purchasing price for electricity generated from renewable energy sources for a specified period of time so as to ensure cost-effectiveness.

> "Since 2007, Thailand's adder program has offered renewable energy producers long-term contracts to sell electricity at attractive rates."

exploit these "clean" energy sources simply need to be developed further and rolled out on a larger scale. A recent World Wildlife Fund (WWF) report said that Thailand could satisfy all of its energy needs from such sources.

To their credit, the kingdom's decision makers have taken note, and Thailand is making concerted efforts to wean the country off its fossil fuel dependency. In doing so, it has distinguished itself as the regional leader in the promotion of renewable energy, embracing clean technology across numerous sectors. Already, the kingdom is reaping numerous benefits for its efforts.

By reducing imports of fossil fuels, Thailand has improved its energy security and trade balance, while developing a rich resource base of renewable energy that has also generated much-needed jobs in rural parts of the country. While developing domestic fossil fuel sources remains a key priority, the government has also created attractive incentives to entice forward-thinking businesses to invest in producing clean energy.

Among Asian nations, Thailand was one of the first to implement a **feed-in tariff**, or "adder" program, incentivizing the development of renewable energy to encourage public participation and boost private sector investment, especially in solar. Since 2007, Thailand's adder program has offered renewable energy producers long-term contracts to sell electricity at attractive rates. For example, solar producers are eligible for subsidies of up to 6.85 baht per kilowatt-hour paid out over 25 years. Companies that generate power through biomass, biogas, hydro, wind, waste energy and solar are all eligible for the adder program.

Additionally, in 2015 Thailand's Board of Investment designated the renewable energy sector a priority industry for development and

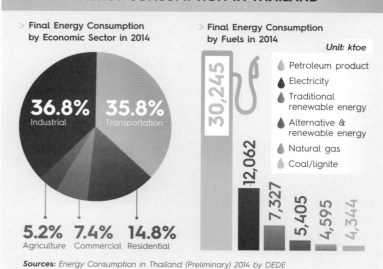

ENERGY CONSUMPTION IN THAILAND

> Final Energy Consumption by Economic Sector in 2014
- 36.8% Industrial
- 35.8% Transportation
- 5.2% Agriculture
- 7.4% Commercial
- 14.8% Residential

> Final Energy Consumption by Fuels in 2014 — Unit: ktoe
- Petroleum product: 30,245
- Electricity: 12,062
- Traditional renewable energy: 7,327
- Alternative & renewable energy: 5,405
- Natural gas: 4,595
- Coal/lignite: 4,344

Sources: Energy Consumption in Thailand (Preliminary) 2014 by DEDE

investment. As such, the government has introduced grants and tax exemptions toward the purchase and import of renewable energy equipment, and it offers foreign investors a number of other incentives to finance renewable energy production in Thailand.

Providing incentives like these to switch to renewables remains important, but alternative energies may soon be able to stand on their own in the Thai marketplace. Carbon credits are already being used to offset the carbon footprints of a range of international organizations and products. Even Thai Airways uses carbon credits from Thailand to offset emissions from their flights, and a small but growing number of Thai businesses offer climate neutral options for their products.

Thailand is also tapping into the full potential of bio-economy-based energy sources. The use of biogas and biomass has already transformed the nature of competitiveness in a number of agro-industries. For example, no starch producer can afford to not produce and use biogas from its wastewater because of its impact on energy-related production costs.

All in all, Thailand is at a pivotal — and exciting — point in its development as a clean energy producer. That said, its reliance on fossil fuels remains firmly entrenched, and with the nation's Power Development Plan (2015) calling for roughly doubling its installed energy capacity by 2036, the ever-growing need for energy has left policymakers with some difficult choices ahead.

For starters, Thailand has a stake in various hydropower projects throughout the region, including the controversial Xayaburi Dam on the Mekong River in Laos, as well as the Hat Gyi Dam and Mong Ton Dam on the Salween River in Myanmar. While on the surface these hold appeal as cheap sources of energy, critics warn that they conceal devastating regional impacts

Greenpeace Mini-Marathon to oppose the coal-fired power plant and seaport in Krabi.

The Cost of Solar Power IS NOW LESS THAN 60 CENTS (21 BAHT) PER WATT

that have yet to be taken into account.

Then there is the issue of coal power plants. The kingdom contains substantial deposits of coal, and lignite has also long been a part of Thailand's energy mix. In 2016, the state-owned enterprise, the Electricity Generation Authority of Thailand (EGAT), confirmed it would construct six new coal-fired power plants by 2025. But scientists, activists and the leaders at the UN see coal power as antithetical to a commitment to sustainable development, given coal's environmental and human impacts and its contribution to climate change. Indeed, plans for two coal-power plants in the southern Thai provinces of Krabi and Songkhla have been met with stiff resistance by the local communities and NGOs concerned about their impact on the region's lifeblood: marine resources and tourism.

Given its significant native energy resources, it is debatable whether the country needs to develop these coal-power plants at all. Some argue that EGAT and other state-owned enterprises shaping these choices are motivated by profit, not the public interest, and wonder how such plans to build coal-power plants are compatible with Thailand's pledge through the Paris Agreement to cut greenhouse gas emissions by 20 percent by 2030. Critics also argue that Thailand should curb demand among businesses and homes for electricity by regulating its usage or offering incentives to cut back, instead of setting higher expectations and targets for production.

One reason Thailand may need to hedge its bets is that its most abundant domestic energy resource, natural gas, which accounts for as much as 45 percent of the country's total energy consumption mix, is projected to dry up in the next decade. Already heavily dependent on gas imports from neighboring Myanmar, Thailand has tasked PTT, the country's state-owned, SET-listed conglomerate to help shepherd the country out of this energy hole. As a part of the solution, PTT and its subsidiaries have aggressively acquired fossil fuel sources abroad such as coal mines and natural gas fields.

THE RISE OF SOLAR AND BIO-ENERGY

It's only natural that an emerging market like Thailand experiences some growing pains, but it's crucial for the nation to pursue sustainable solutions. In the long run, increasing domestic energy efficiency and promoting alternative sources of power offer more promising prospects than a continued reliance on imported energy and fossil fuels. The use of solar energy, in particular, is growing faster than any other power source worldwide. As a country blessed with no shortage of sunshine, Thailand is poised to take advantage. Indeed, solar is fast emerging as the lead alternative energy industry in the kingdom as the technology to cost-effectively harness this clean energy continually improves. Though once astronomical, the cost of solar has plummeted from nearly $75 (2,640 baht) per watt in 1972 to less than 45 cents (16 baht) in 2016. Soon it will be the cheapest form of energy on the planet.

A decade ago solar amounted to less than one-tenth of one percent of Thailand's power. By 2014, it produced about four percent of the nation's electricity or about 1,300 megawatts (MW), a figure that jumped to 2,768 MW by January 2016. Though still modest compared to global solar leaders like Germany and China, Thailand maintains its position as the region's solar powerhouse, producing more than the rest of ASEAN combined. By 2036, Thailand aims to increase solar capacity to nine percent of total electricity generation and

The Energy Conservation Fund

Thailand's Energy Conservation Fund (ENCON Fund) is an innovative way to fund alternative energy projects. Instituted in 1992, the ENCON Fund is sourced from a levy against petroleum products, the rate of which is established by the prime minister. As of 2015, the rate stood at $0.002 per liter. Annually, it generates around seven billion baht, making it capable of significant achievements in sustainable energy development.

Through the levy on petroleum products, the ENCON Fund operates under the "polluter pays" principle. The government can effectively influence fossil fuel pricing (and usage) depending on the levy rate imposed, and guarantee a source of funding for energy efficiency programs. The Energy Conservation Promotion Act requires the ENCON Fund Committee to administer the funds, keeping it separate from the annual government budget allocation system. The Ministry of Energy presides over administrative matters relating to money and fund disbursement. Every five years, the committee establishes a conservation program to serve as a guideline for utilization of the ENCON Fund.

The fund has also helped develop programs like Thailand's Energy Service Company venture capital scheme, the renewable energy feed-in tariff scheme, the provision of tax incentives for energy efficiency projects, and various grant programs. Given its large budget, how the ENCON Fund is managed will likely play a large role in whether the government can achieve its energy efficiency targets under the Energy Efficiency Development Plan (EEDP).

meet the needs of three million households.

It's important to note that Thailand's private sector has been quick to embrace solar, and companies like SPCG are leading the way. As the country's largest solar power firm, SPCG operates 36 solar farms across ten northeastern provinces, which thus far generate a total of 260MW. In April 2016, the government gave the green-light to what will be the kingdom's largest solar energy project to date — some 67 farm cooperatives along with a handful of listed companies have been granted the right to sell a combined 281.32 MW of solar power to the national grid.

A Pa Deng villager feeds manure into his biogas reactor.

> "By 2036, Thailand aims to increase solar capacity to nine percent of total electricity generation and meet the needs of three million households."

Several other renewable technologies are showing great promise as well. Thai Solar Energy Public Company Limited built Southeast Asia's first concentrated solar thermal power plant in Kanchanaburi, and windmills are being tested in various parts of the country, as are small hydro applications.

Biogas, biomass and biofuel also present huge potential, which the country is exploring through a combination of policy support and technological innovation. Using waste to generate power is a logical strategy for sustainable development in Thailand, where a massive agricultural sector creates millions of tons of waste each year. Farms and other waste producers can now sell their refuse as feedstock and generate more profit.

However, even with abundant waste resources, biofuel plants in Thailand face challenges: as higher demand drives the price of feedstock up, biofuel becomes less profitable, the variability in the quality of feedstock

Energy intensity

A measure of the energy efficiency of a nation's economy. It is calculated as units of energy per unit of GDP. Higher energy intensities indicate a higher price or cost of converting energy into GDP. In recent years, Thailand has begun to reduce its energy intensity.

14 Percent
OF THE WORLD'S ENERGY DEMAND IS MET THROUGH BIOMASS

poses operational and cost risks, and the variability in the quality of technology can lower production or increase maintenance costs.

Alternative fuels to power vehicles, on the other hand, face even bigger challenges. Alternative fuels are often mixed with gasoline, making them vulnerable to oil price instability. However, the Thai government is supporting the development of alternative fuels. The Ministry of Energy has set targets to increase Thailand's alternative fuel production. It has also created new feed-in tariffs for energy-crop power production, which are good for 20 years and adjustable with inflation.

Of added benefit, biomass offers opportunities for Thailand beyond the energy sector. With current technological capacities, biomass energy uses only seven to ten percent of the total value of biomass. However, the use of integrated technologies for converting biomass into other materials — such as bioplastics, fertilizers, animal feed and chemicals — can increase this to 70 to 80 percent. Investment in second-generation, integrated bio-refineries is expected to take off on a very large scale down the road, with the potential to turn Thailand into the Middle East of the bio-economy.

The labor-intensive process of turning

Clean Energy:
The Pathway to a Sustainable Future

BIOFUELS

> Biogas, made from the decay of organic matter in the absence of oxygen, is a gaseous fuel high in methane and carbon dioxide. It can be produced from substances such as agricultural waste, manure, plant matter, sewage, food waste, etc.

> Biomass is made of organic matter such as animal waste, rice husks, cassava, palm oil and algae, which can be converted to energy. It can be converted indirectly into energy by distillation into biofuels, such as ethanol and biodiesel. It can also be turned directly into energy by incineration or combustion to produce heat or electricity. Considered one of the key renewable resources of the future, it already supplies 14 percent of the world's primary energy consumption.

> By the beginning of 2014, biomass energy made up the largest share of installed capacity for renewable energy in Thailand at 2,351MW of the total 3,969MW. By 2022 that number is expected to double.

> Biomass energy can be made in ways that help reduce or increase global warming. It can help clean the air, water and soil, or have harmful effects that also threaten biodiversity and bode poorly for public health.

SOLAR AND WIND POWER

> The Renewables 2015 Global Status Report found that 22.8 percent of the world's electricity was generated from renewable sources in 2014. Researchers claim that harnessing the power of the wind and sun would be enough to meet the world's energy demand.

> 2011 marked the first time that global investments in renewable energy surpassed investments in fossil fuels. As of June 2016, clean energy investment had increased to US$286 billion, with solar energy accounting for 56 percent of the total, and wind power for 38 percent. Overall, more than twice as much money was spent on renewables than on coal- and gas-fired power generation (US$130 billion in 2015).

> Power from the sun is the most abundant energy resource. If only 0.1 percent of the energy reaching the Earth could be converted at an efficiency of 10 percent, it would be four times more than the world's electricity generating capacity of 5,000GW.

> Thailand, rich in solar power, generates an average of 18.2 megajoules per square meter per day. (One megajoule can power a 100-watt light bulb for three hours.)

> Solar and wind power are not without their critics. Both sources of power require large tracts of land that could be farmed or reforested. Wind turbines also cause noise pollution and can be harmful to birds.

HYDROGEN AND FUEL CELLS

> Fuel cells, once the stuff of science fiction, now generate pollution-free power from hydrogen and oxygen, leaving only water and waste heat as byproducts. Their use is on the rise, with worldwide sales of fuel cells exceeding US$1.3 billion in 2013.

> Fuel cell efficiencies range between 40 and 60 percent, depending on the type. When waste heat is captured the overall efficiency can be 85 percent.

> Automakers Hyundai, Audi and Toyota have all announced their own brands of hydrogen-heavy Fuel Cell Vehicles (FCVs).

"CLEAN" COAL

> Coal provides 40 percent of the world's electricity while producing almost the same amount of global CO_2 emissions. In Thailand, coal supplied 20 percent of the fuel for power generation while natural gas supplied 67 percent in 2013. Currently, three coal-fired plants generate 4,186MW. To reduce dependence on natural gas, new coal-fired power plants will be built to provide up to 4,400MW more energy by 2030. Facing opposition to this "dirty" energy, the government plans to construct more power plants using a new "clean coal" technology.

HYDROPOWER

> Hydropower provided Thailand with 109MW of power in 2013. By 2022 this figure should surge to 324MW, though there is widespread opposition. As a result, additional power must come from projects that are based in neighboring countries such as Laos and Myanmar.

Government Makes Long-Term Energy Plans

The government's Alternative Energy Development Plan 2015–2036 (AEDP) was drafted with the aim of increasing the supply of alternative energy to meet 30 percent of the country's needs by 2036. Maximizing the potential of alternative sources can be difficult as much depends on the further expansion of the industry through a wide range of supportive measures in terms of incentives, innovation and infrastructure. Some of the key aims of the AEDP are to:

› Promote the installation of small solar projects at the building, household and community level.
› Support the construction of wind turbines and hydro projects in remote areas not currently connected to the national power grid, and in some cases allow local administrations or communities to partly own, manage and maintain the projects.
› Bolster the practice of power generation from waste in small and medium local administrations, schools, temples and communities.
› Encourage biogas generation at the household level and develop community biogas networks, and promote biomass generation owned and managed by community energy enterprises.
› Boost usage of biogas in the transportation sector with pricing mechanisms that reflect actual costs to encourage production.
› Champion ethanol as a gasoline substitute and biodiesel as a diesel substitute; encourage commercial use of other alternative crops such as sweet sorghum.
› Advocate for the use of biomass, biogas, waste and solar to produce heat for the industrial sector, replacing fossil fuels.
› Study the potential to generate tidal energy from domestic sea sources, and develop pilot projects if research deems such further testing worthwhile.
› Identify raw material sources for domestic hydrogen production, and conduct research and development on production technologies and storage.

Thailand's Energy Efficiency Development Plan 2015–2036 (EEDP) outlines several initiatives to help the government achieve significant energy savings, including short- and long-term policy targets, programs and activities, and funding. The primary goal of the EEDP is to reduce Thailand's final energy intensity by 30 percent in 2036 (as compared to 2010). Energy intensity is a way to assess a country's energy efficiency, measured by how much energy is needed to generate a unit of GDP. A high energy intensity may be unsustainable in the future. Financed by the Energy Conservation Promotion Fund, the EEDP prescribes the following five strategic approaches to achieve its target:

› Mandatory requirements such as energy-efficiency labeling on equipment and appliances, and minimum energy performance standards.
› Energy conservation promotion and support such as subsidizing energy savings activities.
› Public awareness campaigns to change consumer behavior. For example, promoting efficient driving practices (idling stop and gentle acceleration/slowdown).
› Promotion of research and technology development, including more energy-efficient and affordable LEDs and electric vehicles.
› Human resource and institutional capacity development to create a local green job market and ensure long-term viability of energy savings programs.

In addition to meeting the target of reducing energy intensity by 25 percent, the EEDP has a long-term target of reducing overall energy consumption by 20 percent from projected business-as-usual levels in 2030 and a short-term target of reducing overall energy consumption by 5 million tons of oil equivalent (mtoe) by 2015. By achieving these targets, the government anticipates reducing final energy consumption by 289,000 kilotons of oil equivalent, avoiding 976 million tons of carbon dioxide emissions, and saving a total of 5.4 trillion baht on energy expenditures. Funding for the program will come from the Energy Conservation Promotion Fund (ENCON Fund).

AFFORDABLE AND CLEAN ENERGY 91

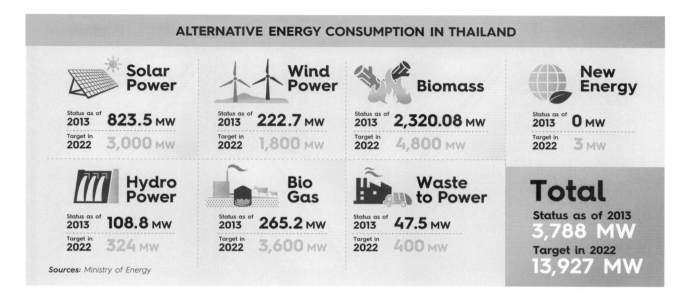

waste into resources also represents a massive economic and labor opportunity, through the creation of jobs and added value. Thus, biomass is not only about power; in the bigger picture these products may go far in displacing Thailand's fossil economy entirely.

Energy crops such as Napier grass are also expected to play a very large role in the mid-term future: unlike agricultural waste, these crops are grown for the explicit purpose of energy production. While Napier grass may be a relatively new energy crop, it's no stranger to Thai farmers as it has been planted locally for more than three decades.

In the short-term, costs are the biggest challenge facing alternative energies because the initial investment is higher in comparison to fossil energy solutions. Alternative energy production systems, technologies and infrastructure to connect to the national power grid all require large upfront investments and can take decades to pay for themselves. Every form of alternative energy also comes with its own set of logistical challenges: solar and wind power require vast tracts of land, while gasohol and ethanol require gas stations capable of distributing the fuel. Further steps can, however, be taken to help promote the proliferation of alternative energies. Technological improvements that increase productivity and lower costs are certain to drive private sector investment. Banks offering green loans can help fill the financing gap for alternative energy projects, especially for small- to medium-sized enterprises (SMEs). Finally, raising public awareness about the cost, health and environmental benefits of clean energy can potentially create a groundswell of demand that shifts the goals and operations on the supply side.

That said, stronger policies and implementation mechanisms are needed for Thailand to achieve energy sustainability and self-sufficiency. The country's success or failure on Goal 7 ultimately boils down to a question of whether or not the government and state-owned enterprises that drive energy policy are truly committed to reshaping the nation's energy mix to fully embrace renewables, or if these bodies will continue to hedge bets by pursuing parallel fossil fuel projects that contribute to climate change and wed the nation to unsustainable energy resources.

Further Reading
• *The Great Transition: Shifting from Fossil Fuels to Solar and Wind Energy* by Lester R. Brown, 2015

• *Thailand: Energy Outlook and the Contribution of JGSEE* by Bundit Fungtammasan, 2014

• *Thailand Energy Report 2015* by Energy Policy and Planning Office, Ministry of Energy

DECENT WORK AND ECONOMIC GROWTH

8 DECENT WORK AND ECONOMIC GROWTH

Promote sustained, inclusive and sustainable economic growth, full and productive employment and decent work for all.

Thailand Looks Toward Its Next Economic Transformation

Calls to Action

- Protect labor rights and ensure safe, secure working environments for all, with a special emphasis on migrants, women and those with disabilities
- Adopt development-oriented policies that promote entrepreneurship and generate quality jobs
- Promote sustainable growth in high-value added sectors through diversification, technology and innovation
- Scale up training in high-tech industries and vocational fields; promote human resource development
- Address the loss of competitiveness in the manufacturing sector due to the emergence of cheaper labor in neighboring countries
- Increase support for SMEs and the informal sector
- Eradicate forced labor and human trafficking
- Generate jobs in tourism that promote local culture and products while safeguarding the environment

An employee wires a Ford EcoSport SUV at the Ford Motor Co. factory in Pluak Daeng, Rayong province.

Here in Thailand, as elsewhere, the model of development that emphasizes GDP as the key measure of progress has come under question and inspired calls for a change in direction toward a more balanced and socially inclusive approach. Most agree that Thailand's economy is at a turning point, and across the board the labor force is in need of an upgrade.

Thailand needs to become an innovator, embrace new technology as it is rolled out and develop top-notch skills training initiatives to meet the growing demand for advanced-level workers in many sectors – from farming to manufacturing. One historic obstacle to this has been that the majority of companies in Thailand have never taken seriously the need to invest in their employees or focus on human resource development. Yet that is exactly what is needed. At present, Thailand is turning out graduates – both from universities and vocational institutions – who are ill-equipped to meet the challenges they face in the modern working world. Improved training in vocational occupations, better English and so-called 21st-century skills such as technology literacy are necessary for Thailand to maximize the potential of its economy and its people. While vast improvements in the education sector are required, employers must also do their part to pick up the slack, and improve their workforce from within.

> "More attention is urgently needed for these informal sector workers, whose work — by the definition of the National Statistical Office — does not offer any social security or protection."
> -Thai Health Working Group

The development of a support structure for informal workers would also go a long way toward ensuring that vulnerable individuals aren't left out in the cold. Another significant topic to address is the lack of adequate protections for migrant workers. Success in this area requires that unethical and unsustainable employment practices come to an end.

In addition, by improving the quality and security of its workforce, Thailand can ensure it is better prepared to pursue a sustainable growth model that embraces high-value added and advanced manufacturing technology.

Due to rising overheads and an increased minimum wage, much of the kingdom's textiles industry has migrated to countries with cheaper wages, such as Cambodia and Bangladesh where birthrates are higher and average ages are younger. Countries with more skilled workforces and higher productivity, such as Vietnam, are also soaking up a growing share of manufacturing investments. These losses are noteworthy given that manufacturing, more than any other sector, drives the Thai economy (accounting for 27 percent of its US$395 billion GDP).

These recent issues aside, Thailand has worked hard to keep its people at work. As of 2015, the workforce consisted of 39.2 million people, or almost 71 percent of the working age population. And the unemployment rate has been on the decline over the past decade. Between 2011 and 2013 it averaged 0.7 percent, the second-lowest rate in the region, thanks in no small part to the massive informal sector. Such positive numbers, as well as the steady price of goods and services, made Thailand the "happiest economy" in the world out of 74 countries surveyed, according to Bloomberg's 2016 Misery Index.

From street vendors to motorcycle taxi drivers to freelance Internet entrepreneurs to small mom-and-pop businesses, Thailand has one of the largest informal economies in the world. Indeed, a whopping 64 percent of Thai workers make ends meet in the largely unregulated, untaxed and often unaccounted for informal sector, also commonly referred to as the "grey economy" — though experts say the actual figure could be much higher.

This sector is a critical source of jobs for the poor or less educated, and for people who

INFORMALLY EMPLOYED

64% of Thai workers operate in the informal sector.

0.7% of the workforce is considered unemployed.

KEY NUMBERS

30% Roughly the percentage of the labor force with university degrees or higher who joined the informal sector. According to an SCB report in 2014, this figure has risen from 22% a decade earlier.

300 US$9 current minimum daily wage.

1.6-3.5 MILLION Migrant Workers

80% of whom are from Myanmar.

Source: Department of Labor, National Statistical Office of Thailand

want to be their own boss. The downside is that informal sector workers rarely receive security or social benefits, and can easily end up exploited by unscrupulous employers.

The country is well aware that these individuals need greater support, as well as formal training so they can improve their lot in life and provide better opportunities for their children. In its 2010 report "Health Indicators of Thailand's Workforce", the Thai Health Working Group, an organization of local health experts, stated: "More attention is urgently needed for these informal sector workers, whose work – by the definition of the National Statistical Office – does not offer any social security or protection."

GIVING SMES THE SUPPORT THEY DESERVE

In Thailand, small businesses are a big deal. In fact, there are some 2.9 million "small- and medium-sized enterprises", or SMEs, scattered across the kingdom. Altogether these outfits make up 99 percent of all enterprises in Thailand and account for nearly four out of five of the nation's jobs. They also contribute 40 percent to the country's annual economic output and 30 percent of export earnings.

Closely related to the presence of SMEs in a country is the level of entrepreneurship among its population, and Thailand scores very highly on this front. The Global Entrepreneurship Monitor found that Thailand exhibits one of the highest entrepreneurship activity rates of the 70 economies it surveyed around the world in 2013. The study found that 46.3 percent of Thailand's adult population is involved in such activities (18.3 percent started or ran "new businesses" and 28 percent identified as "established business owners" – the second highest in the world after Uganda).

Thai SMEs are involved in the full gamut of business activities. About a third are in the service sector including hotels and restaurants, while manufacturing, trade and maintenance make up the bulk of the rest, according to the Office of Small and Medium Enterprises Promotion (OSMEP). Many Thai SMEs operate in the country's booming tourism industry. This sector's vital importance to sustainable development is recognized in Target 8.9, which calls on nations to "devise and implement policies to promote sustainable tourism that creates jobs and promotes local culture and products".

Over the years, Thai governments have recognized the value of SMEs, but it wasn't until the 1997 Asian Financial Crisis knocked a

A laborer works on the roof of a new structure in the popular tourist destination of Khao Lak.

One Tambon, One Product (OTOP)

Inspired by Japan's successful One Village One Product program, OTOP is a local entrepreneurship stimulus program aimed at encouraging tambon (subdistrict) communities to improve the quality and marketing of local products, like foodstuffs and traditional handicrafts.

AT A GLANCE

SMES IN THAILAND
Criteria: "Small" means less than 50 employees while "medium" means 51 to 200

2.9 million	Number of SMEs
99%	Percentage of Total Enterprises
4 out of 5 jobs	Provides Employment
40%	Contribution to Economic Output
30%	Contribution to Exports

hole in Thailand's economic model that anything substantial was done about it on the policy front. The Small and Medium Enterprises Promotion Act B.E. 2543 was passed in 2000 and OSMEP was set up in 2001, with the prime minister or deputy as chair. A year later, the SME Development Bank of Thailand was established. In another crucial development, the Market for Alternative Investments (MAI) – intended to provide a simpler, lower-cost alternative for smaller firms to list on the main board of the Stock Exchange of Thailand (SET) – had its first stock listed in 2001. Another important initiative has focused on upgrading producers in the **One Tambon, One Product (OTOP)** community enterprise program to become SME operators.

But while OSMEP enjoys a certain level of stature, it lacks the authority of a ministry. The 2011 OECD study of Thailand's SME sector

Aspects of Labor in Need of Work

EDUCATION MISMATCH Because the majority of Thai university students prefer to major in social sciences, it remains difficult for Thai industries to find skilled scientists, technicians and engineers.

LACKLUSTER ENGLISH SKILLS Thais have a comparative disadvantage over their regional neighbors in speaking foreign languages. Throw in bad study habits as well as an outdated rote learning system and the result is that Thai workers are considered to be among the poorest English speakers in Southeast Asia.

DOUBLE-EDGED SWORD The "informal sector," comprising freelancers and employees of businesses with fewer than five workers, has been expanding rapidly. This is a double-edged sword that cuts both ways. On the one hand, it has exacerbated the labor shortage in the formal sector. On the other, it has provided a fallback plan for university graduates with degrees in disciplines that possess few career prospects.

WORKERS' RIGHTS WRONGED Labor protection practices are inadequate in Thailand. As much as 75 percent of the total workforce consists of informal workers, many of whom live in poverty, especially those working the land and living upcountry.

LABOR UNIONS Labor unions in Thailand remain relatively weak. In general, they face a number of challenges including anti-union discrimination, denial of the right to organize and restrictions on collective bargaining. Labor unions also tend to be highly fragmented and would be more effective at advocating for workers' rights if they could learn to coordinate better.

SKILLS TRAINING NEEDED Thailand faces impending labor and skills shortages due to an aging population and a workforce lacking proper qualifications in some areas. Deeper emphasis on vocational and industry-specific skills is needed. Small- and medium-sized companies should also offer employees more career development courses and incentives.

AEC WORRIES Deeper regional integration as a result of the AEC could cause the relocation of both investments and industries, as well as a concomitant loss of jobs in Thailand. The key challenge is to combine education and skills development strategies that enhance productivity and competitiveness.

MIGRANT WORKERS MISUSED With barely any bargaining power, migrant workers suffer from unfair wages, low quality of life, rights abuses and the absence of a social safety net or any legal protections.

RETIREMENT PLANNING In anticipation of the aging society that is looming in light of Thailand's low birth rate, the country's public and private sectors need to push for policies that promote saving and motivate people to work beyond the traditional retirement age.

Forced Labor on the High Seas

Migrants sort fish on a Thai fishing boat in Sattahip, Rayong province.

While Thais benefit from strong labor laws, in practice these protections do not extend to migrants, who regularly suffer abuse at the hands of employers or middlemen. Massive numbers of both legal and illegal workers from neighboring countries are predominantly employed in **3D jobs** that are, in general, considered beneath the dignity of most Thai workers. At present, more than 1.5 million migrant workers are registered at the Office of Foreign Workers Administration, but the government and labor experts estimate that the actual figure, including illegal workers, is closer to 3.5 million, some 80 percent from Myanmar.

In recent years, news reports containing credible evidence of slavery and human trafficking in the Thai fishing industry have made headlines abroad and cost Thai businesses millions of dollars. The seafood business, an important economic driver for Thailand, pulls in some US$7 billion per year. According to the Food and Agriculture Organization (FAO), Thailand was the world's third-largest exporter of fisheries products in 2012.

The vast majority of fishermen on Thai trawlers are migrants from Myanmar and Cambodia. A 2011 report by the International Organization for Migration (IOM) called "Trafficking of Fishermen in Thailand" revealed that many of these migrants sought work in Thailand through middlemen. While some got work in food processing factories or in other menial positions, many ended up on long-haul fishing trawlers. Having no contracts and owing huge debts to brokers, the men were sold to the captains or boat owners and forced to work on their vessels for months or years at a time, trawling all the way from Indonesia to India and as far away as Somalia. At sea, many were resold to other captains to prevent the men from escaping back to the mainland and reporting crew leaders to authorities.

The report estimated that some 300,000 migrants are employed in the fishing industry, accounting for 90 percent of the sector's workforce. Some 17 percent can be classified as forced labor. Even though about 50,000 fishing trawlers are registered, there are half that many unlicensed vessels and those lacking proper registration (commonly known as "ghost boats") at sea. For the migrants staffing them, the occupational hazards include 20-hour shifts, beatings, malnutrition and, in extreme cases, execution-style killings.

In theory, migrants are protected by labor laws and regulations that prohibit human trafficking. But in reality existing legislation offers migrants scant protection and no recompense. State agencies overseeing this sector often lack the resources and manpower to stop exploitation and trafficking of migrants, or they collude with the culprits. In 2014, *The Guardian* reported that some Thai authorities were not just lenient but also complicit in such abuses.

The negative publicity from that exposé, and other damning reports by news agencies such as the Associated Press, has put a dent in bilateral relations with the US, which downgraded Thailand to Tier 3 (the lowest ranking) on the State Department's "Trafficking in Persons" index in 2014. In the middle of 2015, the EU followed suit, giving Thailand six months to address the problem of illegal fishing or face a potentially devastating seafood embargo. To its credit, Thailand has shown some improvement in addressing the issue, announcing the implementation of a new system for registering previously undocumented migrant workers that in theory would also protect them from gross injustices and prosecute human traffickers. The kingdom also introduced a system to track fishing vessels. As a result, the US bumped Thailand back up to Tier 2 in 2016, but not without a degree of grumbling from rights monitors.

With its increasing dependence on migrant labor, Thailand needs to carefully calibrate policies to ensure that these foreign workers are treated with dignity. Such safeguards are necessary to avoid more stains on the kingdom's image and to sustain the country's economic growth. According to IOM, the sustainability and profitability of the fisheries sector depends on making systemic changes that will protect nomadic fishermen and promote a fairer, more carefully regulated industry. •

noted, "There is a limit to the ability of OSMEP to achieve policy coherence across government ministries and agencies... [and it] lacks the operational clout and authority to achieve this objective fully."

The current government has vowed to pay greater attention to the needs and contributions of the SME sector, so that it may play a greater role in the country's development. As such, in August 2014 the government allocated a budget of 726.7 million baht to help SMEs improve their products, services and competitiveness. But this figure – even assuming it is deployed effectively – reflects the lowly status of the SME sector compared to other priorities of the state. For example, it is about one-tenth of that allocated in the 2015 national budget for the Ministry of Culture and about a hundredth of the agriculture budget.

In addition to increasing funding for SMEs, the government needs to find incentives to lure them out of the shadows. It could be said that most Thai SMEs are institutionally invisible – only a fifth are registered and of those only half pay any taxes. As the chairman of the Federation of Thai Industries, Supant Mongkolsuthree, said, "If the government wants to change this, it needs to offer special tax rates to lure these businesses to register in the formal system." Indeed, bringing them out of hiding and integrating them more into the visible economy is crucial to formulating good policies and providing guidance.

Those who know the real economic needs of SMEs, such as the Asian Development Bank, point in particular to the lack of finance and the high cost of funds. However, the kingdom is finally taking some measures to address this. In 2015, the Business Collateral Act (BCA) was passed into law and the majority of its provisions came into effect on July 1, 2016. The BCA aims to establish a legal system that allows borrowers in Thailand to use their assets as collateral to secure loans. If effective, the system will provide Thai borrowers with greater access to credit, which should promote investment in domestic ventures.

While the establishment of the BCA shows marked progress, the government could explore further ways that it can encourage the formalization and growth of SMEs. The OECD study makes plain how vitally important this is: "Productive entrepreneurship and innovative and internationalizing SMEs will be key drivers of future economic growth and must be given due attention in policy reform."

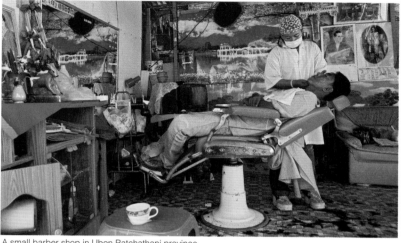

A small barber shop in Ubon Ratchathani province.

Source: *Office of Small and Medium Enterprises Promotion (OSMEP)*

3D jobs
"Dirty, dangerous and difficult" or "dirty, dangerous and demeaning" work, frequently undertaken by migrant laborers instead of the local population.

Further Reading

• *Employment Practices and Working Conditions in Thailand's Fishing Sector* by the ILO and the Asian Research Center for Migration under the Institute of Asian Studies, Chulalongkorn University, 2013

• *Asian Development Outlook 2016: Asia's Potential Growth* by the Asian Development Bank

9 INDUSTRY, INNOVATION AND INFRASTRUCTURE

Build resilient infrastructure, promote sustainable industrialization and foster innovation.

INDUSTRY, INNOVATION AND INFRASTRUCTURE
Sharpening Thailand's Competitive Edge in the Name of Sustainability

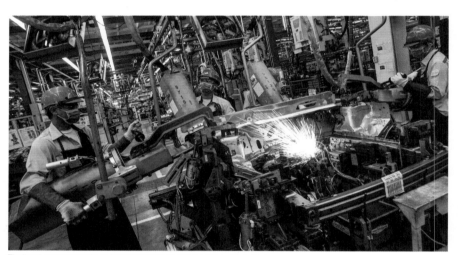

Thailand must upgrade its key industries to move up the value chain.

Calls to Action

- Transition away from low-value manufacturing and export-oriented industrialization toward knowledge-based, high value-added service industries
- Adopt new and advanced technologies more quickly
- Invest more in R&D, innovation and the development of Original Design and Original Brand Manufacturing
- Create more training programs and educational institutions that have better resources and networks
- Encourage more companies to incorporate sustainable development, green manufacturing and CSV principles into their business practices
- Continue to improve connectivity and upgrade transport infrastructure

The development of a country is often synonymous with its industrialization. As an economic engine, a maker of products, an incubator of ideas and an employer of a vast number of people, industry touches all of our lives in direct and indirect ways.

As the conversation about sustainable development spreads, the demand for industrial sustainability standards grows steadily. Although Thailand is already taking steps to meet these standards, this is the bare minimum required for the nation's industry to stay competitive. The real test will lie in making long-term plans for innovation and infrastructure that support growing trade, cultivating a skilled workforce and ensuring political stability that inspires confidence.

In the last century, Thailand was the region's brightest development success story. The nation shot to its leading position among the Newly Industrialized Countries (NICs) in the 1980s and 1990s with a mushrooming manufacturing and subcontracting sector that focused on pumping out cars, garments, electronic components and machinery for global brands. Manufacturing currently accounts for 27 percent of Thailand's roughly US$395 billion GDP, by far its biggest contributor. Thailand is now number 17 in the list of the world's top 20 manufacturing countries.

But today, "the Detroit of Asia" is in danger of losing its foothold as its ability to innovate stagnates and a growing list of cheaper competitors with similar skills rise among the ASEAN nations. Thailand's first stage of industrialization succeeded because it was founded on low-cost, semi-skilled labor and a spirited welcome to foreign investors. But the next stage has proven much more difficult. Thailand is trailing badly in research and development (R&D) spending, the foundation of the value-creating innovation needed to drive productivity, new technology, designs and a competitive "home" brand. According to the World Economic Forum's Global Competitiveness Report 2016–2017, Thailand innovates less than countries with comparable education ratings – investment in R&D has been stuck for decades at just 0.2–0.3 percent of GDP.

This low percentage is one reason Thailand's score in the SDG Index for Goal 9 was its second lowest for any of the Goals. And recognizing a need to do better is why the kingdom is now pushing full-steam-ahead on Thailand 4.0, an economic model based on creativity, innovation and high-level services.

Thailand 4.0 is designed to transform the kingdom into a value-based economy by reforming its major industries (i.e., automotive, electronics, medical and wellness tourism, food, agriculture and biotechnology); scaling up the development of new sectors such as robotics, digital, aviation, logistics, biofuels and biochemicals; and solidifying Thailand as a medical hub.

The blueprint for Thailand 4.0 singles out innovation, entrepreneurship, sustainability, community-led development and inclusive growth as all being essential to its success,

THAILAND'S MANUFACTURING SECTOR AT A GLANCE
(out of 138 countries)

- 16th largest manufacturer of goods
- 60th for R&D investment at 0.2% of GDP
- 24th largest in export volume
- 54th for innovation capability
- 14th most competitive manufacturing nation
- 63rd in technological readiness

Source: Thailand Board of Investment (BOI); World Bank Ease of Doing Business; Global Manufacturing Competitiveness Index; IMD World Competitiveness Yearbook 2013; World Economic Forum Global Competitiveness Report, 2016-2017; The Economist Pocket World in Figures 2016

which is fine in theory. However, the question is: how fast can Thailand's industrial model undergo a makeover?

In the past, Thailand's industrial policies have been sporadic at best and counterproductive at worst. Political and governance failures, lack of continuity, lack of coordination and what might be called the "smile curse" have contributed to an industry sector that has burned bright but is in danger of fizzling out. Like the "resource curse" (the fate suffered by countries that are rich in a natural resource, such as oil, and consequently neglect other, more sustainable aspects of their economy), the "smile curse" is Thailand's welcoming attitude to foreign capital. While it has attracted massive Foreign Direct Investment (FDI) and brought rapid growth, at the same time it has led policymakers to neglect more self-reliant economic options, in particular the development of its own original design and brand manufacturing.

One example is the electronics industry where Thailand has been a global leader in the manufacturing of hard disk drives. But in the past few years, the world has moved on: mobile and other consumer markets are far outpacing the market for such computer hardware, meaning the annual global demand for them has fallen by more than five percent on average since 2011. By comparison, annual global demand for flash memory solid state drives (SSD), in which Thailand has no comparable share, has risen by over 100 percent on average. The decision about whether to make these technologies in Thailand cannot be left to foreign tech companies alone.

Moving up the **value chain** is essential for ensuring that Thailand adapts to such a rapidly evolving world market. Thus far, Thailand's industrial sector has persistently relied on imported capital goods, intermediate inputs

> "The transition from resource-driven, export-led economies to more sustainable growth models based on human capital development, new technology and innovation will be a key challenge for many Asian countries over the next decade."
>
> -Creative Productivity Index: Analysing Creativity and Innovation in Asia, a report by The Economist Intelligence Unit for the Asian Development Bank

and technology, which created a trade deficit that hindered industrial growth. In order to avoid the **middle-income trap**, Thailand's industrial sector must invest in R&D and innovation to cater to changing export markets, new technologies and consumer tastes. Emerging on the horizon, for example, is the digital economy, or economy based on digital technologies and the Internet of Things. This is perhaps the most rapidly growing sector in the world, as well as one of the most important drivers of innovation, competitiveness and growth. Many other countries, not just Thailand, have been slow to implement the logistical framework and know-how to grow this

Value chain
The successive steps through which a product is made, from purchase and delivery of the raw material, to manufacturing operations and distribution, marketing and sales, with each stage generating revenue for the company providing the service. Different stages of the chain can be performed by different companies. The "higher up" the chain (design, innovation, application of advanced technology), the more revenue that can be earned.

Middle-income trap
Some development economists say most countries hit a growth ceiling after a few decades of rapid expansion because they fail to move "up the value chain". This means they continue to compete on low-cost labor and low-quality goods rather than improve the skills and productivity of the country's workforce and infrastructure.

Thailand must improve its research and development capabilities.

sector at home. In fact, according to the European Commission, 41 percent of European enterprises are non-digital, and only two percent are fully taking advantage of digital opportunities. This, coming from one of the largest economies of the world, seems to imply that it is not too late to get in the game. But Thailand must act swiftly to establish itself as an integral player in this lucrative field.

Another way Thai industry can offer added value to the world market is to incorporate the idea of "creating shared value", or CSV. The main principle behind CSV is to pair business growth with long-term positive environmental and social impacts. In today's increasingly sustainability-oriented marketplace, CSV comes at a premium – more and more consumers and investors are willing to pay higher prices for the products and services of companies that can prove their sustainability. Heavy industries are also recognizing that their unsustainable business practices directly affect the environment, place burdens on society and ultimately add to the cost of business itself. Carbon pricing is a good

HOW A MAJOR CONGLOMERATE IS USING SEP TO FOSTER INNOVATION

SCG is one of Thailand's oldest and leading conglomerates with operations throughout ASEAN. The company has been instrumental in the development of Thailand's infrastructure and industry. However, its core products — paper and packaging, cement, logistics and petrochemicals — are associated with high environmental impacts. In 2007, SCG announced its vision to become a sustainable business leader and has since integrated the concepts of sustainable development throughout its value chain with a focus on strong corporate governance, human resources, R&D, zero-waste, efficient energy use, eco-products and more. At the heart of this push toward sustainability is an incremental approach toward expansion and a desire to promote innovation in its working culture and product lines so that SCG can continue to grow sustainably and help the country prosper. The company's business strategy also adheres to SEP's key principles and dimensions. Below is an overview of how SCG's practices help promote sustainable development:

Economy
- Efficient utilization of resources
- Product innovation
- Increased income

Society
- Vibrant communities and healthy employees
- Safe cities

Progress with balance

Environment
- Environmentally friendly processes
- Responsible production & services
- Clean energy

Culture
- Strong company values and ethics
- Strong community network

Moderation
- Challenge investors to take a long-term perspective by paying stable dividends while maintaining share price
- Invest long-term in cultural development, management, society, environmental conservation, and the quality of products and services, even though such investments may temporarily reduce profits
- Restructure and reduce the number of business ventures, and only expand those that represent the core strengths of SCG

Reasonableness
- Promote people from within the company
- Ensure that all employees are utilized and at their full potential
- Implement the 3Rs (Reduce, Reuse, Replenish) in business operations

Prudence
- Plan succession for top management, including the CEO
- Seek to gradually expand to other markets overseas in a carefully planned manner
- Encourage employee participation through One Cell One Project
- Invest heavily in employee development

Knowledge
- Run programs to promote employee creativity
- Build networks to share knowledge
- Focus on building R&D capabilities and incremental innovation throughout the organization

Virtues
- Integrity and fairness to all stakeholders is one of the core values successfully cultivated through recruitment, training, promotion, and leadership role modelling
- Sponsors numerous long-term environmentally and socially beneficial projects

Source: *Thailand Sustainable Development Foundation, 2016*

Cleaning Up Transport's Act

High transport costs are lowering productivity and profits in Asia's export-oriented economies. Costing between 15 and 25 percent of GDP, they are two to three times higher than in the US, Europe and Japan. The environmental impacts are also staggering. Trucks account for less than ten percent of the total number of vehicles in Asia but are responsible for about half of all the CO_2 and particulate matter emitted by the transport sector.

In response, major companies are greening their supply chains to reduce costs, mitigate delivery risks and improve brand image. Government and international organizations are also building platforms and developing policies to push the sector toward achieving greater energy efficiency. On the recently established Green Freight Asia Network, members share methodologies for data collection and analysis, best practices and performance scorecards. It helps industry heavyweights like IKEA, DHL and UPS improve the CO_2 and fuel efficiency performances of their own fleets as well as their subcontractors.

In the US, the SmartWay program helps truck operators compare technologies, like aerodynamic trailers or wire-based tires, and secure financing for desired improvements. Launched in 2004, the program now covers 650,000 trucks (or 30 percent of US road freight), working with 2,900 shippers and carriers.

Thailand is taking steps to improve the efficiency of its logistics operations too. For example, the Logistics and Transport Management Application (LTMA) project, jointly developed by the Federation of Thai Industries and the Energy Policy and Planning Office, enhances logistic and transport management through technological advancements. In total, 104 companies with 5,396 trucks updated their operational practices by the completion of the first of three phases in May 2013.

---

example of this: the more CO_2 a company emits, the more it hits their bottom line.

At the end of the day, Thai industry simply cannot afford to be unsustainable and slow to innovate any longer. "If the country moves toward modern farming and knowledge-based, high value-added service industries, it will be able to escape the 'middle income trap'," said Somkiat Tangkitvanich, president of Thailand Development Research Institute.

SOLUTIONS IN INFRASTRUCTURE AND INNOVATION

The solutions for how to grow sustainable industries lie in resilient infrastructure and relevant innovation, the two main pillars of industry. While infrastructure provides the transport, education, policies and logistical aspects of building an industry, innovation offers the new ideas, science and technology that improve industries and set the economic agenda. Let's look at infrastructure first.

For the most part, Thailand has relatively good foundations to support industry, with a large workforce, good roads and ports for import-export, and policies and bodies (such as the Board of Investment) that promote industry and trade. The World Bank's 2016 Ease of Doing Business Report places Thailand at number 49 among 189 countries. In the 2016–2017 Global Competitiveness Report, Thailand ranks 34 among 138 nations, with a particularly strong showing in its macroeconomic environment.

However, improvements can always be made, especially now with the rise of the Asian Economic Community (AEC). With Thailand's geographic location in the middle of the ASEAN member nations, the country is situated to become the logistical hub of Southeast Asia, making transport infrastruc-

Connectivity

In economics this refers to linking supply chains (the production, shipment, assembly and distribution of various product components) through better roads, rail, ports and airports as well as more efficient customs procedures.

ture and "**connectivity**" all the more important for economic growth. Transport also has a direct effect on the country's competitiveness, with more direct and efficient ways to move goods playing a key role in consumer choice.

That's one reason Thailand is ramping up its infrastructure investment, following more than two decades of inadequate spending in this key area. The kingdom is on track to spend some US$3 billion on infrastructure development in 2016, a figure that is expected to rise to an average of US$9 billion per year until 2020, according to Maybank Kim Eng. The focus will be on constructing new railways, roads and customs posts to establish cross-border trade routes and thus improve Thailand's overall quality of rail infrastructure and import-export facilities.

The most ambitious project currently on the table is the 352-kilometer Bangkok-Nakhon Ratchasima double-track high-speed-railway. Originally to be a jointly funded venture with China, the project – estimated to cost 170- to 190-billion baht to complete – is now to be wholly funded by Thailand, with China providing construction support. Meanwhile, the government is also working to expand the country's highway network by 20 percent by building some 2,650 kilometers of new roads.

Inefficient transport infrastructure also takes a major toll on national wealth. In 2013, Thailand's transportation costs were equivalent to 7.4 percent of the gross domestic product of the country, while its logistic costs accounted for 14 percent of GDP. This is not to mention the carbon footprint and external costs of the transport sector. In 2013 for example, transportation alone was responsible for 25.2 percent of all CO_2 emissions from energy consumption in Thailand.

Also significant are the logistical requirements for supporting the rapidly growing digital economy. Getting the people, technologies and networks that can help conduct digital businesses are a part of good infrastructure too. Both the private sector and the government must partner to put in place high-capacity broadband Internet, data centers and verification systems to promote the cyber security needed for e-commerce. In fact, the Board of Investment is already moving in this direction: as of 2015 it began offering incentives for investors to support high technology and innovation projects.

So how does innovation then play a role? Simply put, investment in innovation creates the thinkers, engineers, technologies and ideas that give rise to higher-value, cutting-edge, sustainable industries. And this is what Thailand sorely lacks.

The Thai government and private sector

"The World Bank's Ease of Doing Business Report places Thailand at number 49 among 189 countries."

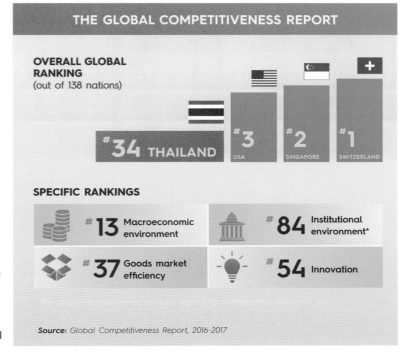

Source: Global Competitiveness Report, 2016-2017

Integrating "Sufficiency Thinking" in Business

Sustainable development initiatives in Thailand's private sector are often inspired by Sufficiency Economy Philosophy (SEP) principles. Most such actions were taken after the 1997 Asian Financial Crisis when Thailand realized it needed to strengthen its immunity to risks. King Bhumibol Adulyadej's SEP provided a perfect framework to apply to both small and large companies.

Government and non-government organizations have also sought to support SEP's integration into corporate policy, value chains and governance through contests that reward companies for their best practices or through the creation of new industry standards firms can achieve. The Office of the Royal Development Projects Board (RDPB) organizes a national contest to identify SEP businesses. The winners were appointed as learning centers on how to run an "SEP business". The first-time winners among large, medium and small businesses were SCG, Chumporn Cabana Resort and Nithi Foods, respectively. The second-time winners were Bangchak Petroleum, Bathroom Design and Porntip Phuket, respectively.

The RDPB, Thailand Research Fund, the Thai Chamber of Commerce and the Board of Trade of Thailand have also supported Mahidol University to develop the "Sufficiency Economy Business Standard", which is now being used by agencies such as the RDPB, the Thailand Sustainable Development Foundation and Thai Credit Guarantee Corporation as a tool to promote the adoption of SEP among business people. It is a standard available at no charge.

The Ministry of Industry has introduced an Industrial Standard No. 9999 to Thai industries. It is developed based on the SEP with an aim to promote sustainability in the nation. Meanwhile, the Office of the National Economic and Social Development Board has established a network of large Thai companies such as SCG, PTT Group, Bangchak Petroleum and Toshiba to implement SEP with their trade partners throughout their value chains. All of these actions are helping to promote not only King Bhumibol's ideas but also the principles of sustainable development. The success of such practices among some companies is dispelling the myth that maximizing short-term shareholder value should be the top priority.

thus far have not put adequate investment into R&D and innovation, including the educational institutions, resources and networks that are required to create a highly skilled workforce. In fact, the World Economic Forum's Global Competitive Report for 2016–2017 shows that Thailand's ranking for innovation has fallen to 54 (out of 138 countries) from 33 in 2007. Despite years of discussion about the need for improvements to the education system, successive governments have yet to address the core problems – namely, a focus on rote learning and standardized testing verses the need for critical thinking and the development of 21st-century skills.

Cultivating innovation may require challenging, systemic changes, but concrete steps can be taken. To start, ensuring that the government has up-to-date knowledge on emerging markets and industry trends will only go toward supporting the entrepreneurs and investors who want to transform Thailand's low-value manufacturing model into the more profitable Original Design or Original Brand Manufacturing industries. Some companies like SCG are already pushing Thai industry toward the next level of innovation in sustainable production. In 2009, the company launched its environmentally friendly product label called "SCG Eco Value." The company's 84 SCG Eco Value products range from marine cement, paper and bathroom tiles to biodegradable plastic pellets. And such forward-thinking practices are paying dividends. In

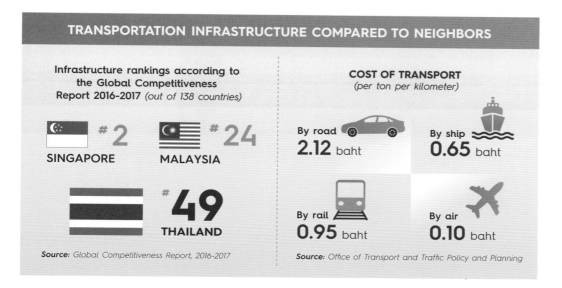

2014, SCG's Eco Value products generated 31 percent of the company's revenue.

Financial institutions also have a key role to play in this field. Some Thai banks such as KASIKORNBANK already offer green loans to businesses that are planning environmental or energy reduction retrofits. This can be expanded to include infrastructure projects to help make transport and logistics more sustainable, while low-interest loans for R&D projects can help incentivize innovation efforts. In addition, the entire value chain of Thai industries can be made more sustainable by providing microloans to help rural entrepreneurs and small-scale suppliers. Oftentimes, the greatest change can come from the bottom up. The rise of a more sustainable value chain, workforce, suppliers, entrepreneurs and innovators can move mountains – or in this case, change industries.

Finally, the government must recognize that far more than well-meaning policies are required to foster competitive industries. The bedrock for sustainable growth is political stability. Look, for example, at the hit that FDI in Thailand took after the May 2014 coup d'état. As investor confidence fell after the military takeover, investment from the EU plummeted from 86.7 billion baht in 2014 to just two billion baht in 2015. These kinds of drops in investment can seriously hinder development, not least of which because much FDI in Thailand goes toward infrastructure projects.

While political stability may remain largely in the hands of lawmakers, the government can take a more active role in encouraging the private sector to invest in innovation, to integrate King Bhumibol Adulyadej's Sufficiency Economy Philosophy and to move toward CSV in their corporate policies and value chains. Creating incentives, developing business standards and offering rewarding contests are all ways to push the private sector on a more sustainable trajectory. However, until Thailand can show the world a stable political system, even the best projects and policies for sustainable industry, infrastructure and innovation will only continue to be stalled and hindered.

Further Reading

• *Thailand: Industrialization and Economic Catch-Up* by the Asian Development Bank, 2015

• *The Impact of Finance on the Performance of Thai Manufacturing Small and Medium-Sized Enterprises* by the Asian Development Bank, 2016

10 REDUCED INEQUALITIES

\ Reduce inequality within and among countries.

REDUCED INEQUALITIES
Vast Divides Separate Thailand's Haves and Have-nots

Calls to Action

- Implement and enforce more egalitarian, meaningful taxes on land, wealth and capital gains
- Develop better social welfare programs for the elderly and disadvantaged
- Promote income growth in the bottom 40 percent of the population at a rate higher than the national average
- Encourage social, economic and political inclusion
- Achieve more equal distribution of public resources
- Adopt fiscal, wage and social protection policies that promote equality
- Empower and promote the social, economic and political inclusion of all, irrespective of age, sex, disability, race, ethnicity, origin, religion or economic or other status

In Thailand, power is concentrated in the hands of the few, and the distribution of wealth, land and resources remains highly imbalanced when compared to other upper-middle-income countries.

The issue of inequality is shared between countries rich and poor. From the United States to Thailand, income inequality, rural-urban divides and the unequal distribution of public resources are hot button issues inciting contemporary debate. So it is not surprising that the entire 2030 Agenda for Sustainable Development resonates with the urgent call to reduce inequalities: in terms of income and quality of education, access to justice and to services such as healthcare, and between nations themselves.

Policymakers from around the world have acknowledged that inequality is a potential threat to long-term social and economic development and stability. The solution, it is argued, is to flesh out more inclusive development policies that involve all stakeholders and better balance the three dimensions of sustainable development: economic progress, social welfare and environmental impacts.

Reducing inequalities is an incredible challenge when power and wealth are being accrued more and more by the few: the richest ten percent of earners are taking home up to 40 percent of global income, while the poorest ten percent earn just two to seven percent, according to the UN. As of March 2016, the world had 1,810 billionaires, who together hold some US$6.5 trillion in wealth, according to *Forbes*.

Thailand is very enmeshed in this contemporary conundrum, though with some very Thai-specific features. While some of the kingdom's richest tycoons are themselves self-made billionaires, their rags-to-riches stories are outliers that belie the pervading sense that Thailand is not distributing its opportunities fairly. Yes, economic growth has helped millions of Thais out of poverty, but as of 2016, the distribution of wealth and resources remains highly imbalanced when compared to other upper-middle-income countries.

Once an absolute monarchy, where the king essentially owned or controlled the majority of capital, Thailand is still a place where power and wealth remain remote to the vast majority of the population and the sense of participating in a civil society is weak. While Thailand certainly has expanded its middle class, decision-making and resources are still concentrated in the capital, and the divides between Bangkok and the rest of Thailand, between policymakers and communities, and between the owners of capital and the vast workforce doing their bidding have not been meaningfully bridged. After decades of imbalanced growth, some critics now frame Thailand not as a growing democracy, but as an oligarchy run by a network of elites that revolve around the monarchy, top companies, state-owned enterprises and other power centers.

It should come as no surprise that many Thai leaders and observers see the current situation as increasingly untenable. "For development to be sustainable, the fruits of economic growth must be spread widely and fairly to foster social cohesion and continued economic and political legitimacy," said former prime minister Anand Panyarachun in a 2016 speech to the Foreign Correspondents Club of Thailand. "Many of the economic and social problems we currently face, including the simmering political tensions and sporadic clashes we have suffered in the past decade, can be traced back to the injustice and inequality inherent in our society."

INCOME, LAND AND TAXES

What are some of the key sources of inequality in Thailand? To begin, Thailand suffers severe income inequality, which is reflected in its score as calculated by the Gini coefficient. This accepted measure calculates the extent to which the distribution of income among individuals or households within an economy deviates from a perfectly equal system distribution. A low Gini coefficient indicates a more equal distribution, with 0 corresponding to complete equality, while a higher coefficient indicates more income disparity. By this internationally recognized standard, Thailand's level of income inequality has fallen from 0.52 in 2000 to 0.47 in 2013. This figure, however, is still higher than Indonesia, Laos, Vietnam and Cambodia and means Thailand has among the highest income inequality levels in Southeast Asia. This fact and others led Thailand to score below its region's average on Goal 10, according to the SDG Index.

Thailand also features a radically unequal distribution of land. The top ten percent of all landholders (roughly 1.5 million individuals and juristic organizations) hold more than 60 percent of all land, while the bottom ten percent hold just 0.7 percent, according to data from the Land Department. The largest landholder, Thai Beverage chairman Charoen Sirivadhanabhakdi, reportedly holds some 101,000 hectares. Meanwhile, in 2013 members of parliament reported holding land that, on average, was worth almost 31 million baht

Gini coefficient

Devised by the Italian statistician Corrado Gini, the Gini coefficient measures the extent to which the distribution of income among individuals or households within an economy deviates from a perfectly equal system distribution. A low Gini coefficient indicates a more equal distribution, with 0 corresponding to complete equality, while a higher coefficient indicates more income disparity.

(about US$900,000). These numbers speak to some of the vast inequalities at play in Thailand and the need for more effective measures to better balance land and wealth distribution – something the kingdom continues to struggle to enshrine in its governing laws.

Since the 1990s, a slew of regressive tax measures have fueled an environment whereby Thailand's rich can gain from the tax system. "One cause of the great inequality in wealth distribution is that Thai governments have never seriously taxed land or assets, but instead have often offered tax benefits that actively contribute to greater inequality", wrote Sarinee Achavanuntakul, Nathasit Rakkiattiwong and Wanicha Direkudomsak in their paper "Inequality, the Capital Market and Political Stocks". As an example, they pointed to a 2007 initiative that temporarily increased the amount of income exempted from taxation if the income was invested in a Retirement Mutual Fund or Long-term Equity Fund. Even though only some 11,000 people with annual net incomes of at least four million baht could benefit from the measure, the exemption amount was increased from 500,000 to 700,000 baht.

The issue is that Thailand's income tax system is simply "not designed to be fair", wrote Pan Ananapibut, head of the Fiscal Policy Office's Tax Structure Development subdivision, in his paper "Tax Reform for a More Equal Society". "The system facilitates tax evasion by the rich and thus contradicts the principles of fair taxation and ability to pay. Many of the tax exemptions and allowances result in low-income people subsidizing the high-income ones", Pan wrote.

"The fundamental reason for the low level of public goods is a low level of taxation", wrote Pasuk Phongpaichit in a 2012 article published by the *East Asia Forum Quarterly*. In Thailand, the ratio of tax to GDP is around 17 percent. In many other middle-income countries with similar levels of development, the ratio is much higher: 25 percent in Venezuela and 32 percent in Turkey. "Taxation is not only low, but it may also add marginally to inequality by weighing more heavily on the poor than the rich", Pasuk added.

According to Duangmanee Laovakul, an Assistant Professor in the Faculty of Economics at Thammasat University, a land tax, wealth tax or tax on capital gains that yields revenue that can be spent on increasing the supply of public goods and resources would "have a positive impact on inequality".

This theory, of course, is based on the assumption that monies collected from taxes would then be distributed and used in an efficient manner that supports the public good. But given the corruption common to past Thai governments and Thai bureaucracy in general, as well as the ineffectiveness of many state projects, such tax revenues would not necessarily translate into a significant impact.

To date, even sincere government efforts to address land and wealth redistribution through policy change have gained little ground. In 2011, the National Reform Commis-

> "One cause of the great inequality in wealth distribution is that Thai governments have never seriously taxed land or assets, but instead have often offered tax benefits that actively contribute to greater inequality."
>
> -Sarinee Achavanuntakul, Nathasit Rakkiattiwong and Wanicha Direkudomsak from "Inequality, the Capital Market and Political Stocks"

Thailand Ranks 89th Worldwide IN TERMS OF THE GINI COEFFICIENT MEASURE OF INCOME INEQUALITY

sion proposed wide-ranging land reforms including a progressive tax on people owning large tracts of land, a cap on private land ownership, a proposal to allocate land to a million poor families and a system to provide legal assistance to villagers facing encroachment charges. But five years on, success is uncertain. A proposed inheritance tax of five percent has been revised by the National Legislative Assembly (NLA), which raised the minimum value subject to inheritance tax to 100 million baht. Meanwhile, a tax bill on new land and buildings, which had received the green light from the cabinet, was altered to be applied to properties worth 50 million baht and over.

UNOFFICIAL OLIGARCHY?

The constant protection and advancement of the wealthy, whether through favoritism in the form of tax breaks or cronyism, has led to a contentious debate about the structure of Thai society as a whole. As Pasuk Phongpaichit and Chris Baker wrote in their 2016 book *Unequal Thailand: Aspects of Income, Wealth and Power*, "informal networks and coalitions of the few play major roles in the distribution of power and economic benefits..." They add that "these economic inequalities underlie inequalities of power, social position and access to resources of all kinds." The Director of the Asia Research Centre at Murdoch University, Kevin Hewison, argues a similar premise in a 2015 article in the *Kyoto Review of Southeast Asia*: "The inequality of conditions in Thailand is the fundamental fact from which all others are derived. Economic and political inequalities in Thailand are mutually reinforcing conditions that have resulted from the ways in which the gains of rapid economic growth have been captured by elites."

Many Thailand observers would agree and argue that dating back to the era of absolute monarchy through to present day, the "fruits of economic growth" continue to be shared only among a few. Instead of building an accountable and transparent meritocracy based on the rule of law, these critics argue that the country's elite have used their elevated status to stack the odds in their favor and protect the various apparatus that further their financial ambitions and solidify their power base. "Relatively low incomes, skewed ownership and the siphoning of income to the already rich indicate a long-standing pattern of exploitation", Hewison said.

According to this criticism, in many ways, maintaining "the gap" has become almost as important as the profits, monopolies and alliances it helps to engender. Pasuk and Baker argue that mechanisms have been put into place by elites to safeguard the very existence of Thailand's equality gap. "The persistence of economic inequality is a function of the strength of oligarchy. The rule of the

At the Thai-Cambodian border, itinerant traders pass by an ad for easy credit.

REDUCED INEQUALITIES

Youth from economically disadvantaged households grow up having far fewer opportunities than those from well-off families.

few is found not only at the national level but also in the operation of institutions at all levels of society. The few rule and prosper by cultivating and defending privileges and monopolies of various kinds, and by opposing

> "Economic and political inequalities in Thailand are mutually reinforcing conditions that have resulted from the ways in which the gains of rapid economic growth have been captured by elites."
> -Kevin Hewison, Director of the Asia Research Centre at Murdoch University

extension of the rule of law which might form the foundation for greater equality", Pasuk and Baker write. As related to Target 10.2, which promotes the "political inclusion of all," the feeling that the powerful public and private sectors do not properly listen to the concerns of community stakeholders has fueled the sense of inequality in Thai society.

As an example, Nopanun Wannathepsakul points to the hybrid semi-public, semi-private nature of PTT and the Electricity Generating Authority of Thailand. "These massive hybrid organizations in the energy sector have been created by a 'network bureaucracy,' which commands great official power. This network spreads across all agencies involved with the energy sector...The hybrid nature of these corporate groups and the power of this 'network bureaucracy have contributed significantly to the rapid expansion of these corporations," Nopanun wrote in the 2016 paper "Network Bureaucracy and Public-Private Firms in Thailand's Energy Sector".

At the same time, Nopanun says that some executives in these organizations "hold or once held high public office; they are chairpersons or directors on several boards; at the same time they have high positions in public agencies that oversee these organizations...Their remuneration from these multiple posts is generous."

Decentralization, Anand reasons, would enable "the participation of more diverse interest groups and represents one way to curb the concentration of power and influence exercised by political forces." The influence of connections, nepotism and pure greed of those in power has not gone unnoticed by the average Thai, with the result that inequality, and the structures that maintain it, are increasingly being challenged through political protests.

THE OPPORTUNITY DIVIDE

Indeed, the concentration of decision-making, wealth and income in Bangkok has, in turn, spawned an urban-rural divide in terms of access to resources, services and opportunities at all levels. Duangmanee said "one

Thailand's Social Progress Ranking IS 57 OUT OF 133 COUNTRIES

Anti-coal banners in Krabi's Nua Khlong district. Local communities often feel that vested state and corporate interests are overriding the genuine concerns of average citizens.

reason for inequality in Thailand is the under-supply of public goods and services" reflecting a long-held complaint of those outside Bangkok that resources are too centralized. Indeed, in provincial Thailand everything from educational resources to physical infrastructure to hospital equipment to public funding is lacking when compared to the country's thriving capital.

"There are too few good schools, inadequate public transport facilities, and no comprehensive provision for old age even though the society is rapidly ageing," Duangmanee wrote in her paper "Concentration of Land and Other Wealth in Thailand". Thus, for the poorest of the poor, especially those living outside Bangkok, the pathway to a decent or higher education, a quality job, justice or simply a bank loan can be difficult to imagine, let alone access. Darker skin, family name, birthplace, gender and other indicators outside of an individual's control are often implicitly considered during, for example, job hiring. This, in turns, creates a "cycle of deprivation" in which a family's lack of opportunities are transmitted from parents to children and onto grandchildren, ensuring that future generations remain trapped by the status quo.

For such people, access to quality education is a prime example. Principals of elite schools are known to expect fees, donations or outright bribes to admit students and acceptance is largely based on one's ability to pay tuition. Children of the elite can afford to attend private Catholic or international schools, hire tutors or travel abroad to learn foreign languages or other vital skills, and therefore have a significantly better chance to follow in their parents' footsteps. Meanwhile, rural children who primarily attend under-funded public schools lack qualified teachers and other essential education resources. The simple fact is that poor-quality education during primary and secondary years hinders the ability of a child to later pursue higher education and unleash his or her potential.

"Youth from socially and economically disadvantaged households have fewer opportunities to enter tertiary education than youth from privileged households", creating a pattern of lower wages when the former enters the labor force, said Dilaka Lathapipat of the Thailand Development Research Institute, in the paper "Inequality in Education and Wages". Thailand needs to tackle the "challenging task of overcoming wealth-related inequality in college preparedness from an early age by providing good child care facilities in poor communities and eliminating the huge disparity in the quality of basic education provided by resource-poor and resource-rich schools," Dilaka said in a 2012 paper, "The

Thailand's "Village Fund"

Thai technocrats have long called for municipal and provincial governments to have greater say over state investment in local projects and the direction in which they are carried out. In an effort to decentralize spending, in 2001 the Thai government launched the ambitious Village and Urban Community Fund, which allocates discretionary funds to more than 79,000 communities nationwide. Fund allocations and activities are determined by a village committee, which while operating with autonomy, strives to create a local development plan that encourages holistic activities, emphasizes shared knowledge, cooperation and development that is in tune with each village's capacity. The fund is intended to help both individuals and villages build resilience through balanced community-led development. The fund's micro-credit program lets borrowers take out small, low-interest loans without putting up collateral, thus serving as a valuable mechanism for improving equality. To date, some 200 billion baht has been dispersed through the fund and is in circulation.

Influence of Family Wealth on the Educational Attainments of Youths in Thailand".

Many argue that creating more equal access to education resources is also essential to national unity. "The educational and opportunity gap must be reduced in order for Thais to build trust among each other," Bank of Thailand Governor Veerathai Santiprabhob's said in a speech in September 2016.

While government program after government program has failed to adequately resolve the issue of inequality, it is not necessarily for lack of effort. The state-owned Bank of Agriculture and Agriculture Cooperatives (BACC), established in 1966, provides loans to more than 5.6 million farm households and has over 1,200 branches nationwide. From the close relationships it has forged with the farmers, it has proven itself to be knowledgeable and responsive to their needs. Indeed, the issue of inequality has been on the agenda of nearly every government dating back to the early 1960s era of Field Marshal Sarit Thanarat, who wanted to help pull the farmers of the northeast out of extreme poverty. Technocrats and government ministers, such as Kosit Panpiemras, dedicated much of their careers to trying to improve the lot of farmers. The royally initiated projects of King Bhumibol Adulyadej have been predominantly focused on creating security and opportunities for the rural poor, creating better healthcare in provincial areas, and educational and work opportunities. Yet the struggle to tackle inequality persists.

What is needed is a truly integrated approach that focuses on creating a more inclusive society. And it's not about making the rich poorer, but rather, it's about ensuring that progress and development occur in a more balanced, fair manner. "We have to build an environment that supports dynamic change…. The environment must not be limited by outdated rules or rules that benefit any particular group of people that does not want any change", Veerathai said.

One solution would be the sincere application of the principles advocated by SEP and sustainable development in general: the idea that economic, environmental, social and cultural concerns all need to be considered. If local communities are empowered to control more of their own resources, if wealth and land are distributed in a more reasonable manner, and the rich and powerful act with more moderation and listen to the wishes of the people, a reduction in inequalities would likely result. Likewise, if Thailand can improve its performance on other key Goals – such as improving innovation in industry, promoting decent work, creating empowerment opportunities for the disadvantaged and ensuring more equal access to justice – a corresponding reduction in inequality should follow.

A worker walks in front of an advertisement for a luxury condo in Bangkok. Due to their backgrounds, many Thais are largely shut out from the opportunities and prosperity found in Bangkok.

Further Reading

• *Unequal Thailand: Aspects of Income, Wealth and Power* by Pasuk Phongpaichit and Chris Baker, 2016

• *Inequality and Politics in Thailand* by Kevin Hewison, the Kyoto Review of Southeast Asia, 2015

• *Property Tax in Thailand: An Assessment and Policy Implications* by Duangmanee Laovakul, 2016

• "Democratic Governance: Striving for Thailand's New Normal," a speech by Anand Panyarachun to the FCCT in March 2016, available online

11 SUSTAINABLE CITIES AND COMMUNITIES

Make cities and human settlements inclusive, safe, resilient and sustainable.

SUSTAINABLE CITIES AND COMMUNITIES

Toward Cleaner, Greener and More Inclusive Urban Development

Calls to Action

- Find sustainable ways to address rapid urbanization and urban equity
- Continue to promote more equitable access to physical and social infrastructure including affordable serviced land, housing, energy, water, sanitation, waste disposal and mobility
- Upgrade more slums through inclusive development
- Improve road safety
- Create more urban green spaces and parks
- Improve urban air quality and waste management
- Support positive economic, social, logistical and environmental urban-rural linkages
- Implement integrated policies and plans that address inclusion, resource efficiency, mitigation and adaptation to climate change and resilience to disasters
- Promote the construction of sustainable and resilient buildings utilizing local materials

People walk across Chong Nonsi pedestrian bridge in Bangkok, which is home to some 8.8 million people.

Roughly half the world's population resides in cities today, and by 2050, two out of every three people will be urban dwellers. Over the next few decades, 95 percent of urban expansion will take place in the developing world. And while cities occupy just three percent of all land, they account for 60 to 80 percent of energy consumption and 70 percent of greenhouse gas emissions.

With such figures, the challenge of maintaining cities in a way that continues to promote quality jobs, adequate shelter, a healthy environment, solid infrastructure and inclusive prosperity, while not straining resources, creating pollution or causing human suffering, looms large in many countries around the world, especially those with enormous populations.

The rapid growth of cities in the developing world, coupled with increasing rural to urban migration, has led to a boom in megacities. In 1990, there were ten megacities with 10 million inhabitants or more. In 2014, there are 29 megacities, home to a total of 453 million people. Extreme poverty is often concentrated in urban spaces, and national and city governments struggle to accommodate the rising population in these areas. Making cities safe and sustainable means ensuring access to safe and affordable housing, and upgrading slum settlements. It also involves investment in public transport, creating green public spaces, and improving urban planning and management in a way that is both participatory and inclusive.

> "Improving the quality of data to understand trends in urban expansion is important, so that policy makers can make better-informed decisions to support sustainable communities in a rapidly changing environment, with access to services, jobs and housing."
>
> -Marisela Montoliu Munoz, director of the World Bank Group Social, Urban, Rural and Resilience Global Practice

A popular Thai saying claims that Bangkok is Thailand. Indeed, as the hub of the kingdom's politics, commerce, media and much more, the capital – while not technically a "**megacity**" – is what is known as a "primate city", dominating over the rest of the country. With no other city boasting more than 500,000 residents, Bangkok accounts for 80 percent of Thailand's total urban area, according to the World Bank, and is the fifth-largest urban area in East Asia. The capital's glitz and prosperity may be envied upcountry, but the drawbacks of this urbanization – the crowds, the pollution and the traffic – are just as notorious.

As the city expands – it's now home, temporary or permanent, to nearly 8.8 million people – Bangkok has also become the country's capital of consumption. A third of the electricity generated in Thailand goes to feed the power-hungry needs of urbanites, with shopping malls alone devouring more electricity than all of Cambodia and Siam Paragon complex using twice as much electricity as Mae Hong Son province. Dams built to satisfy Bangkok's electricity needs have also depleted fish stocks in Laos. Moreover, Bangkok's carbon dioxide emissions per person are more than ten times those of a Northeasterner and the city consumes more water than the rest of the country combined. Much of the city's water has to be pumped in from upcountry reservoirs. Because of a lack of wastewater treatment plants, the discharge (mostly untreated) goes straight back into the Chao Phraya River and various canals.

Air pollution is another perpetual hazard. While Bangkok's air quality has improved and ranks better than many other major cities, the country needs to reduce the concentration of its fine particulate matter (or "PM2.5") in order to meet Goal 11, according to the SDG Index. In a 2013 study, the Pollution Control Department (PCD) claimed that Bangkok has the third-worst air quality in Thailand, ranked behind the district of Na Pralarn in Saraburi province, home to the country's biggest gypsum factories, and the Map Ta Phut Industrial Estate on the Eastern Seaboard. To achieve Goal 11, Thailand also need to increase access to improved water sources from 76 percent of the population to 98 percent. The city's sanitation capacity has also been pushed to capacity. The excess refuse must be dumped at landfill sites in nearby provinces, such as Samut Prakan, where locals have protested after big fires at garbage dumps left an acrid, eye-watering stench in the air for weeks.

All these symptoms are suggestive of poor urban planning. Instead of expanding northwards, as city planners have recommended, Bangkok's mid-section continues to widen,

Megacity

A megacity is usually defined as a metropolitan area with a total population in excess of ten million people. A megacity can be a single metropolitan area or two or more metropolitan areas that converge.

Thailand's Sustainable Cities Progress

The UN has identified four main pillars required to build a sustainable city: social development, economic development, environmental management and urban governance. Although the world has 29 megacities, the World Bank estimates that almost half of the world's urban population lives in settlements with less than 500,000 citizens. This model of what is often referred to as the compact city has become the focus of Thailand's sustainable city projects.

The concept of sustainable cities was introduced in Thailand in 2004 by the Ministry of Natural Resources and Environment and is based, in part, on the Sufficiency Economy Philosophy of King Bhumibol Adulyadej, which can be applied not only to rural development but also to urban development. Thailand's Sustainable City project intends to strengthen local governments in environmental management by promoting public participation, developing capacity-building activities and transferring experiences from city to city. Project implementation started with 15 local governments as pilot sites and 485 local governments collaborating in the "Sustainable City Network". Some of the most successful Sustainable City Projects have been awarded by the ASEAN Working Group on Environmentally Sustainable Cities (ESC).

Phuket city, for example, has developed several strategies to encourage the development of a sustainable, healthy city. With the collaboration of the private sector, government agencies and communities, Phuket city implemented an air quality monitoring system that measures key pollution factors to ensure ideal standards of air quality. In collaboration with the Traffic Police, Pollution Control Department and Land Transport Department, roadside inspections are carried out habitually to encourage the regular maintenance of automobiles for emissions reduction. To maintain good air quality and to simultaneously reduce solid waste accumulation, Phuket city has designed the "Waste to Energy Program". This program produced 1.7 megawatts of electricity from solid waste incineration, accounting for 2.7 percent of total electricity consumed in the city. In order to achieve better long-term results, environmental activities have been promoted through public information campaigns, which encouraged active participation in carbon emission reduction and better energy consumption in homes as well.

In Thailand's North, to become an ESC, Chiang Rai city initiated "clean and green land programs" emphasizing

Phuket has implemented several sustainable development initiatives.

solid-waste management through the **3R's** and increasing green areas. The program outlines comprehensive strategies for efficient solid-waste management, including creating a cooperative network between private waste collectors, municipal waste collectors and private recyclers to encourage waste separation; creating recycling banks among schools and different neighborhoods; developing school programs to teach the 3R's and to encourage activities where students transform solid waste into useful materials; setting up marketplaces to sell products made of recycled materials; developing waste collection with a fleet of trucks and volunteers who report any uncollected solid waste; and the improvement of a five-pit sanitary landfill by creating three wastewater treatment ponds to control leachates as well as seven groundwater-monitoring wells to prevent polluting leaks.

The city also developed a Green Area Management program to increase green space, conserve and protect biodiversity, maintain water balance and improve air quality. Chiang Rai has 41,462,100 square meters of green area and 17 public parks. To conserve this green space, Chiang Rai organized a community tree conservation activity where residents created a database of large plant species to increase people's awareness of their environment and to prevent deforestation. Chiang Rai's green efforts have been recognized internationally.

spilling over into its eastern fringes as urbanites follow the expansion of the Skytrain. Condos, housing estates and shopping complexes spring up in their wake. This blueprint of urban development is being replicated across the country. Urban encroachment in floodplains and agricultural areas in Nakhon Ratchasima, for example, was blamed for exacerbating the epic floods of 2011. Up in Chiang Mai, still without a popular public transport system, traffic jams are a growing problem and air pollution is rising to the forefront of ecological issues.

All these cities have essentially emulated the **ribbon development** style. This model, pioneered in the UK in the 1920s and 1930s, spawned both urban sprawl and suburbs as people acquired cars that enabled them to live farther away from the city center. As a result, land prices in the middle of the city shoot up beyond the means of the majority, so the middle classes and poor have to move to the city's outer limits and suffer through long commutes in heavy traffic. This is also why Bangkok has such an excess of motorcycles and cars. Of the 34.5 million private automobiles registered in Thailand almost a quarter of them are in Bangkok. But the city has far too little road area – only ten percent compared to 23 percent in Tokyo or 38 percent in New York City – to handle the number of vehicles.

There are some bright spots on the horizon. In 1999, the Skytrain, or BTS system, got online after more than 20 years of plans derailed by successive governments and corruption. The BTS was followed by the MRT, or subway, five years later. Both of those mass-transit systems, which offer commuters the chance to rise above or duck under the city's gridlock for short intervals, are being expanded now to cover the city's outer reaches. With little room left for ground-level development across much of Bangkok, there is nowhere to go but up. The Skytrain's shaded, open concourses known as skywalks, for example, are now favored over cluttered and damaged sidewalks.

As long as the capital remains the biggest city of opportunity, political power is centralized and material benefits and decent jobs are kept within city limits, it is likely that this slapdash style of urbanization will prevail.

A PLATFORM FOR GROWTH

As Thailand's largest urban center, Bangkok holds significant gravitational pull over the rural poor who flock to the capital in search of better jobs, brighter lights and higher living standards. They form the backbone of the city's underclass and constantly increase the ranks of urban poor living in some 1,000 slums across the city.

In general, slum areas have typically been passed over by governments for improvements, further exacerbating Thailand's growing divide between rich and poor and creating a society of haves and have-nots. Instead of hope and opportunity, the slums can breed

Ribbon development

This term, which came into vogue in the 1920s, describes the pattern of developing houses, buildings and infrastructure projects along transportation routes. It is most often cited as the cause of unregulated urban sprawl.

3R's

Refers to three terms often used when talking about waste: Reduce, Reuse and Recycle.

The BTS Skytrain and other public transportation projects are key to reducing pollution.

crime, violence and substance abuse. Most are overcrowded firetraps with hovels and tin shanties for rooms, which offer little in the way of safety or security for slum dwellers. Many of the residents are squatters with few rights who can be evicted at any time on short notice.

"Creating new urban environments must involve developing communities in a humane and sustainable manner", said Somsook Boonyabancha, a preeminent housing rights advocate with decades of experience in urban planning. "Urban planning should be a process through which experts engage people in collective decision-making."

Somsook is one of the masterminds behind the government-run Community Organization Development Institute's (CODI) Baan Mankong ("Secure Housing") Collective Housing program. Taking a bottom-up approach that helps create a practical vision for communities according to their own needs, the program assists the capital's urban poor in upgrading their living environments within slums through infrastructure subsidies and housing loans. Working hand-in-hand with municipal authorities, experts, urban planners and non-governmental organizations, enterprising slum dwellers can take the initiative in collaborative efforts to make their communities more livable through the creation of better

The Road to a Cleaner, Happier Bangkok

 CAPPING CAR EMISSIONS Promoting cleaner fuels in addition to raising the taxes on older vehicles responsible for a higher level of emissions could help to clear the air.

 BETTER ZONING Zoning laws designed to keep factories and other commercial developments from encroaching on communities need to be enforced.

 RUSTY PIPES Many water pipes in Bangkok need to be fixed or replaced because they have a leakage rate of around 35 percent, well above the average 22 percent rate of the 22 Asian cities included on the Asian Green Index 2011.

 TREATING WASTEWATER The Bangkok Metropolitan Authority (BMA) needs to build more wastewater treatment plants and enforce existing laws on the discharges from factories.

 CUTTING ENERGY CONSUMPTION Mount campaigns to encourage energy conservation through using clean energy sources like solar panels, while discouraging people from using too much air-conditioning.

 GREEN LUNGS Provide tax incentives for building owners who preserve big trees and create new green spaces when they develop their projects.

 URBAN PLANNING Extending the expiration date of town plans from five years to ten years or more would allow for more strategic long-term planning and better ensure the plan's sustainability. It is also of crucial importance to prevent any further revision or amendments to the plans that allow for questionable projects to be built during their implementation period.

housing and public spaces. They can also receive legal and technical support.

Set up in 2003, the agency has helped locals build almost 100,000 new homes in more than 1,800 communities as part of some 933 collaborative grassroots projects. These projects are active not only in Bangkok but also nationwide.

"Proper housing is extremely important for the sustainable development of any urban area", says Somsook. For this bottom-up rather than top-down form of development to really work, the program demands that participants take a proactive approach.

"Our main objective is to empower poor people to deal with their own problems", explains Thipparat Noppaladarom, CODI's former director. "People have to be the principal actors behind sustainable development projects by taking charge and ownership of their communities. In the past, many poor people were content to be passive recipients of help from the government in a top-down approach. Now more and more of them are seeking to help themselves through bottom-up initiatives."

"We sorely need land reform in this country", adds Somsook. "We need to capture the energy of new human potential. All sustainable development models need to be based on our greatest asset: people."

A PROPER ROADMAP NEEDED

In looking for the underlying causes to the unique issues facing Bangkok, such as too few parks and too many cars, policymakers and activists have typically fingered the same culprit: town planning or a lack thereof. Basically a blueprint for how a city should be developed, these master plans are drafted in bigger Western countries or Asian nations like

Low income housing is common along Bangkok canals.

Japan and China years before projects are constructed. But in Thailand the town plan is not completed until after the projects are up and running, if there is any such plan at all.

That makes it easy for corrupt businessmen and politicians to get in on the ground floor of such developments, said Srisuwan Janya, who works as a pro bono lawyer for communities affected by environmental woes. In 2014, Srisuwan helped residents in Soi Ruamrudee in Bangkok win a case against the developers of a high-rise building, because the developers had not constructed the building according to the town plan.

This is commonplace, he said. "Development policies in Thailand are carried out in piecemeal efforts. Many development plans are there not because they should be, but because landowners and investors want them to be there. Roads are built to follow in the path of real estate speculators and those influential people who will get benefits from the construction. Then we have many buildings and infrastructure projects built in the wrong environment, which leads to various problems such as traffic congestion and floods."

A lack of planning and overdevelopment

CODI HAS HELPED BUILD More Than 100,000 Homes SINCE 2003

Source: Thai Climate Justice Working Group

are the causes of many environmental woes in Bangkok, where roads block drains, swamps and canals have been filled in for commercial properties and zoning laws are often ignored. It's not that the authorities have never paid any attention to urban planning (as far back as 1935 the government first attempted to draft a town plan), it's just that any such attempts to draw them up rarely go according to plan. In fact, the first town plan for Bangkok was not drawn up until 1960 by an American firm. The few town plans in existence today expire after five years. And if a new plan has not been put in place then developers can apply to construct new projects with little in the way of scrutiny.

Across the country, only nine provinces have drawn up master plans to allocate land and aid development, according to Ratchatin Sayamanond, former director-general of the Department of Public Works and Town and Country Planning. The kingdom's other 68 provinces have no such plans in place and have shown little interest in drawing any up in the near future.

The WHO's Recommended Standard OF URBAN GREEN SPACE PER CAPITA IS 16 SQUARE METERS

GREENER URBAN PLANNING AND CULTURAL PRESERVATION

Space, proper shading, cultural preservation and comfortable connectivity are important to a city's sustainability. They not only impact the physical and mental health of urban residents, but also promote economic productivity as well as a sense of community. Bangkokians long for the outdoors – and the fresh air, peace and open space it offers.

Unfortunately, few city residents in the world enjoy less public green space than those in Bangkok. The City of Angels has only 5.46 meters of green space per person, making it one of the worst endowed metropolises in the region. The WHO's recommended standard of urban green space per capita is 16 square meters. By comparison, Kuala Lumpur has 12 square meters of green space per capita, while Singapore residents enjoy 66 square meters of green space per person, thanks to decades of proper city planning, which has been largely absent in freewheeling Bangkok.

Campaigning to Save Big Trees

A group of concerned citizens have banded together to protect Bangkok's big trees by working from the grassroots up. Oraya Sutabutr, the founder of BIG TREES, spoke for many of the city's denizens when she said, "No one can deny that the weather has become hotter, the air more polluted and the traffic even worse despite mass transit systems like the Skytrain and subway. People can no longer put up with traffic jams, but they start walking and they realize that the streets are too polluted and too hot, because of the lack of trees and green areas."

One of the most popular and active civic groups in the city, BIG TREES started in 2010 when they launched a campaign to save a huge rain tree in Sukhumvit Soi 35. The tree, located on private property, was going to be cut down as the landowner wanted to sell the plot to a nearby shopping complex. Using social media sites like Facebook, the group united nature lovers at loggerheads with Bangkok's mall-heavy development schemes.

The campaign could not save that tree, but it did succeed in sowing the seeds of awareness about the value of such trees, which give off oxygen, suck up the CO_2 produced by burning fossil fuels and provide a natural form of air-conditioning for houses surrounded by them. Parks and trees tend to boost property prices, the group noted. In Bangkok, big real estate developers such as Sansiri have begun to preserve the original trees around their condos and create green spaces, because they attract high-end buyers and add value to the units.

Since that first headline-grabbing campaign, the group's social media sites have become popular places to congregate for nature-loving people to share their campaigns to protect large trees in their community, or save green spaces in the public domain from being bulldozed for commercial ventures.

A few years ago, BIG TREES spread its roots. They joined forces with an association of architects, architecture students and other civic groups to ask the State Railway of Thailand to convert a large tract of greenery in the middle of the city into a park instead of selling it to developers. That is an ongoing campaign. More recently, the group championed a campaign to prevent the authorities from revising a town plan that would turn the city's "green lung" in Bang Krachao into another industrialized suburb of Bangkok. Through big campaigns like these, the group has garnered more and more advocates, including some high-profile supporters such as Anand Panyarachun, a former prime minister.

Now BIG TREES has joined hands with the Bangkok Municipal Administration's professional arborists to conduct workshops that train laypeople how to take care of trees. Its other efforts have also yielded positive results. In 2014, the group lobbied CentralWorld to spare the trees that blocked access to one of its exits. In the end, the department agreed to design a new exit to save the trees.

Oraya believes that Bangkok, despite its many environmental challenges, is far from a lost cause. When she started doing volunteer work two decades ago, her friends told her she was out of her mind because the situation was hopeless. "Now I see more civic groups and individuals come to help. Bangkok residents do not sit idle any longer, waiting for the authorities to do things for them. They realize that they cannot wait for them and they start to help themselves," said Oraya.

Poor planning and zoning as well as endless construction projects have gradually denuded the city of its greenery and led much of its once world-famous network of canals to be filled in, exacerbating the city's **urban heat island** effect. Air, noise and nocturnal light pollution have also taken their toll on the capital's livability.

Having recognized the need for more green space, the BMA has gradually been turning back the clock, with 980 rai of green space added in 2013, an additional 1,459 rai in 2014, and another 1,350 set to open in 2016, including nine public parks.

In addition to more concern being shown for the protection of trees and the creation of green spaces, the building of eco-buildings is also gradually gaining ground in Thailand. Since 2007, when Thailand earned its first **LEED** certification for a manufacturing plant of the multinational carpet maker InterfaceFLOR, green buildings have been cropping up. Such structures have been commissioned by myriad institutions, from schools (International School Bangkok's Cultural Center) and foreign governments (USAID's Regional Development Mission for Asia) to Thailand's blue-chip companies (PTT's Energy Complex and SCG's 100th Year Building).

These buildings are designed (inside and out) to operate on key sustainability principles, including energy efficiency, reduction of waste and greenhouse gas emissions and water conservation. They are built with careful consideration of materials and the local environment, while also keeping in mind the need for a comfortable working atmosphere.

Today, about a hundred certifiably green structures dot the country, from a Toyota showroom in Nakhon Ratchasima province to a 7-Eleven outlet in Bangkok, to a six-story KASIKORNBANK Learning Center in Chachoengsao province, as well as the Thai Health Promotion Foundation Office. Partly driving the trend are companies branding themselves as eco-friendly. When a business like SCG, whose core businesses have environmental impacts, sells "green products", their headquarters becomes a part of their brand and corporate image.

Meanwhile, just as the internationally recognized LEED certification program by the United States' Green Building Council has been gathering momentum in Thailand, the kingdom's own Thailand Green Building Institute (TGBI) is offering a domestic equivalent for eco-friendly certification: the TREES system. Since the TGBI was founded in 2009, 16 buildings have been certified by TREES.

Green building practices are increasingly critical when you consider that malls, commercial buildings and apartments are among the three biggest energy users in the business sector, according to the Thai Climate Justice Working Group. When you combine all of these

---

Urban heat island

An urban heat island (UHI) is when a city is significantly warmer than the surrounding rural areas, owing to the heat generated by its buildings, infrastructure, vehicles and human activity.

LEED, or Leadership in Energy and Environmental Design

Though not always seen as applying rigorous standards, LEED is the world's leading certification program for green buildings. To be LEED certified, buildings must meet the criteria laid out by the US Green Building Council.

> "Urbanization does not produce only infrastructure. Real urbanization must create active citizens who are not afraid to speak up, who spend time monitoring the authorities and making sure officials do their jobs. It is not real urbanization if the city is fully populated by passive citizens who just complain about problems, but sit idle and let authorities and developers do as they wish."
>
> -Srisuwan Janya, activist and founder of the Stop Global Warming Association

A man rides a bike in Sri Nakhon Khuean Khan Park in Bang Krachao, which is known as Bangkok's "Green Lung".

in Thailand, they consume more energy per year than the entire nations of Cambodia and Laos put together. Working at optimum efficiency, green buildings use up to 50 percent less energy and 60 percent less potable water. They produce 70 percent less solid wastes and emit 35 percent less CO_2. During construction, companies also save money by cutting waste by up to 80 percent. Whether starting small by refitting an office with energy-saving light bulbs or designing a whole new building, the money saved on water and electricity can be significant over the long term.

In Asian countries such as Thailand, McGraw Hill Construction estimates that companies with green buildings can save around 21 percent on operating costs over five years with new buildings, and 13 percent on buildings renovated to be more eco-friendly. This means that the return on investment for such structures is about seven years, faster than in many Western countries where construction costs are far higher. Over time, these savings will defray the exorbitant costs of constructing such monuments to sustainability. SCG's 100th Year Building, for example, set the company back 3.3 billion baht, much more than a normal building of this size. But saving on energy and costs are not the only values represented by buildings.

For Thailand, its historic architecture, often obscured by newer buildings or torn down altogether, represent cultural capital that has often been neglected. In the spirit of conserving this craftsmanship, since 1997 the Crown Property Bureau (CPB) has been preserving or restoring palaces, residences, shophouses and public buildings.

Many of these are not just museums of the past but fully functioning buildings with tenants. This can pose its own challenges. For example, in the case of 22 wooden row houses near the tourism area of Khao San Road, the renovations needed to be completed not only to restore and preserve their aesthetic value but in a way that would not disrupt the livelihoods of the travel agencies, clothing and other shops that were long-term renters. By doing the renovations as efficiently as possible, paying for 75 percent of the costs and creating a new lease scheme that was fair and affordable, the CPB and the tenants managed to cooperate on what might otherwise prove to be a contentious process.

Further Reading
• *Employment Practices and Working Conditions in Thailand's Fishing Sector* by ILO and the Asian Research Center for Migration under the Institute of Asian Studies, Chulalongkorn University, 2013

• *Asian Development Outlook 2016: Asia's Potential Growth* by the Asian Development Bank

12 RESPONSIBLE CONSUMPTION AND PRODUCTION

\ Ensure sustainable consumption and production patterns.

RESPONSIBLE CONSUMPTION AND PRODUCTION

Creating Conscientious Buyers and Ethical Businesses

Calls to Action

- Raise awareness about the benefits of responsible consumption and cultivate mindsets that are more focused on moderation
- Encourage individuals, companies and government agencies to practice more responsible management of waste, natural resources and chemicals
- Incentivize the use of green manufacturing practices and ethical sourcing to make production systems more sustainable
- Curb unethical practices in the production supply chain including human rights abuses and environmental degradation
- Promote sustainable public procurement practices in accordance with national policies and priorities
- Increase conservation efforts in areas despoiled due to excessive tourism; develop strict environmental protection regulations

A pile of broken and discarded mobile phones at a shop in Bangkok.

Humanity is living way beyond the Earth's means, using up 1.5 times more resources than the planet can replenish in a year, as compared to only five decades ago when we were consuming just a third of that figure. Meanwhile, the ever-expanding global population has topped 7.4 billion and is growing at a rate of about 80 million people per year. The ominous reality we're confronted with is that even if we manage to conserve resources through improved efficiency and reduced per-capita consumption, those gains will be offset by this massive surge in human numbers, threatening the planet's continued ability to meet our basic needs. In short, drastic measures must be taken to alter consumption habits, change mentalities and overhaul our production systems before it's too late. And the three stakeholders that drive consumption – individuals, governments and the private sector – will each need to do their part.

In the decades to come, Asia will be at the center of this global shopping spree. It's already home to 40 percent of the so-called "consumer class" – the some 1.7 billion people with disposable incomes worldwide. In Thailand, the Westernization of society has strikingly altered the way this disposable income is spent. Where once owning items like a TV, smart phone or motorbike were a luxury, now they are almost a rite of passage and Thai consumers from all walks of life increasingly want trendy clothes, brand name gear and high-tech gadgets.

However, there is rising awareness in Thailand about the need for more sustainable lifestyles and business practices, and a number of state and private organizations are employing King Bhumibol Adulyadej's Sufficiency Economy Philosophy (SEP) to spread awareness about this issue. Proponents of SEP reason that adhering to its principles of moderation and reasonableness can help individuals be more responsible consumers and encourage businesses to implement sustainable production processes.

With the end goal being to promote sustainability, such notions ring true on a number of levels. Better-informed, environmentally-conscious consumers are more likely to make ethical, knowledge-based decisions about what they buy, in what quantity, and from whom. They recycle, avoid wasteful practices, try to generate less trash, and come up with do-it-yourself answers to stretch resources. Meanwhile, manufacturers that incorporate SEP into their businesses are attempting to be more conscientious about the impacts their processes may have on the environment. They strive to reduce their carbon footprint, use ethically sourced materials and prioritize sustainability over maximizing profits.

However, at least for the time being, modern life in Thailand still revolves – to an extent – around possessions, and this materialistic mentality has led to unsustainable purchasing behavior.

For example, Thais simply love their cars and many are willing to go into debt to obtain the status bump and independence afforded by a new vehicle. In 2011–2012, more than 1.2 million Thais took advantage of an ill-conceived government incentive scheme that offered first-time car buyers a tax rebate of up to 100,000 baht. But, because many of those who took advantage of the offer were from lower-income households, roughly ten percent eventually defaulted on their loans, resulting in the seizure of their cars by finance companies, according to IHS Global Automotive. It also put more than a million new cars on the already congested streets and cost Thailand $2.5 billion in tax breaks, according to the World Bank.

Another example of irresponsible consumption in Thailand is the frivolous distribu-

THAI WASTELAND

1.15 kg per day
The average waste produced by each Thai

61 million per day
Styrofoam boxes used by Thais

3.9 billion per year
The number of plastic bottles of water consumed by Thais

Up to 75%
Amount of household waste that can be recycled

81%
of the total 2,490 waste disposal sites nationwide are considered substandard

Of the almost **27 million tons** of household waste produced in Thailand
64% of that rubbish was food

Source: Pollution Control Department, Food and Agriculture Organization of the United Nations (FAO)

tion of disposable plastic items. Unsustainable practices at convenience stores and supermarkets like 7-Eleven, Family Mart, Tops, Villa Market, Big C and Tesco perpetuate the problem as clerks regularly wrap purchases in several layers of plastic and distribute unnecessary straws and single-use cutlery. Across the country, food stalls and restaurants are guilty of serving up take-away in plastic and Styrofoam containers, too. But convenience obsessed consumers also deserve a hefty portion of the blame, and the notion that a purchase isn't complete until it has been encased in several layers of plastic is one that needs to be addressed on both sides of the checkout counter.

Going Green 101

So many dilemmas related to the environment and society – such as pollution, climate change and excessive consumption – seem so vast and insurmountable as to be completely beyond our influence. But that's not true. Through awareness and moderation of our own personal consumption, the individual can always make a difference. In fact, every consumer choice we make – from what type of disinfectant we use to what light bulb we purchase – has an effect on both the environment and the bottom line of the firms who produce these products.

Here are a few ways that individuals can lead greener, healthier lifestyles and benefit from significant cost savings:

> Say no to plastic bags, straws and spoons. Buy products that come in biodegradable packaging or natural packages like banana leaf.
> Choose LED light bulbs over incandescent or even CFLs. When purchasing electrical appliances, select those that have a level 5 energy efficiency certification.
> Unplug TV sets, computers and other machines when they are not in use to save electricity.
> Wash/dry only full loads of clothes instead of using the laundry machine for only a few garments at a time.
> Surround your house with shady trees and plants to keep it cooler and reduce the need for air-conditioning.
> Grow an organic vegetable garden.
> Use eco-friendly, biodegradable cleaners instead of chemical varieties. Baking soda, for instance, is an effective all-purpose cleaner. White vinegar and borax are good cleaning agents while citrus and tea tree oils serve as natural disinfectants. Meanwhile, damp microfiber cloths, which attract dirt, can do wonders even without any cleaning agents.

> Compost your organic waste instead of throwing it out. Fruit and vegetable peelings, tea leaves and eggshells can become excellent soil enhancers. Likewise, grass cuttings on freshly mown lawns, if left there to decompose, work as natural fertilizer for new grass.
> Ditch single-use paper and plastic products in favor of durable goods like ceramics to reduce unnecessary waste. Likewise, choose recycled products when it comes to items like stationery and toilet paper.
> Buy second-hand books, furniture and clothes. Donate items like old clothes, books, toys and mobile phones instead of throwing them in the trash.
> Divide up glass, paper, metal and plastic trash before putting it out in the garbage. Oftentimes these items are picked up by scavengers who recycle them.
> Avoid taking motorized transport. Instead, walk or ride a bike.

Coffee beans dry at Doi Chaang in Chiang Rai province.

One way to curb overuse is to charge customers a fee if they want to use a plastic bag to carry their purchases. In early 2016, Indonesia launched a pilot program in 22 cities to test the effectiveness of such a tariff on reducing waste. Indeed taxes, rewards and other incentives could be integrated in multiple areas to encourage individuals to practice more responsible consumption. Many of Thailand's 68 million citizens don't realize that their personal choices about how and what they buy do in fact make a difference in the greater scheme of things.

If Thailand wants to achieve the consumer-side obligations of Goal 12, it needs to find ways to make its people care about such details and help them to develop sustainability-focused mindsets. That requires more direct engagement, raising awareness, proactive measures to alter consumption patterns, the promotion of sustainable habits like recycling or reuse, and encouraging people to simply consume less and "go green".

PRIVATE SECTOR INVOLVEMENT IS KEY

A sustainable world is only a Utopian fantasy without the cooperation of the private sector, which holds considerable sway over consumption and production patterns, both in Thailand and across the globe.

Consider this: a single company can have the impact of a thousand households through greener manufacturing practices, investment in energy efficiency, sensitive labor practices, better waste management, or by maintaining procurement policies that reward companies who also believe in responsible production. In many ways, it's about achieving a reasonable balance between progress and sustainability. Toshiba, one of the leading electronic manufacturers in the world, has built a green factory in Thailand that has cut down emissions by 75 percent. Mitr Phol, one of the largest sugar producers in the world, has revolutionized waste-to-energy technology and implementation, while amply demonstrating the profitability of such innovations.

Responsible production is relatively easy to implement within smaller companies as well. The Doi Chaang Coffee Company, which sprouted from humble roots in Chiang Rai province, is an excellent example. Arabica coffee trees, which can grow under the canopy of shade trees and therefore don't require land to be completely cleared, were introduced through a royal project in the early 1970s as a cash crop to help ethnic Akha villagers replace plums, which had seen falling demand. Fast forward to the present, and the brand is now internationally recognized for its high quality Arabica coffee. Because it is produced through sustainable methods that are eco-friendly, provide livelihoods and also help preserve the Akha's indigenous culture, the product also holds great appeal for environmentally conscious consumers.

Such small, ethical businesses, from independently owned restaurants in Bangkok to eco-tour operators on Koh Tao, are now influencing the way the public views consump-

Ethical sourcing
When products are sourced, manufactured and supplied without exploiting people or the environment. Ethically sourced products are created in safe facilities by workers who are treated well and paid fair wages to work legal hours.

Plan Toys: Growing a Successful Brand from Discarded Wood

Plan Toys are made from sustainable materials, glues and dyes.

At the core of Goal 12 is a belief that the means to achieve responsible levels of consumption and production are all around us. We just have to look for them. Since the early 1980s, innovative companies like Thailand's Plan Toys have been doing just that.

Until just a few decades ago, the extensive rubber plantations of Thailand, Malaysia and Indonesia, which produce about 70 percent of the worlds' latex rubber, were simply burned off at the end of their productive life so new rubber trees could be planted. In this wasteful practice, the creators of Plan Toys saw an opportunity and developed a way to exploit the true value of rubber wood as a sustainable raw material.

Based in the rubber-producing region of Trang in southern Thailand, Plan Toys has earned a worldwide reputation for creating imaginative and child-safe toys made from sustainable materials – principally mature, recycled rubber wood.

The factory started production in 1981 and now consumes about 1,000 tons of rubber wood a year. It produces around 5 million toys each year, mainly for export to the United States and Japan, as well as over 60 other countries, and boasts sales of more than US$16 million per year. At the heart of the company's operations are principles of environmental and social sustainability, as well as a highly developed approach to green manufacturing.

Plan Toys uses a chemical-free kiln process to treat rubber wood, thereby protecting it from fungal and insect damage. This key innovation has helped turn what was once disposable rubbish into a sustainable resource and valuable raw material. Once treated, the dense rubber wood is tough and does not split or splinter. Its natural colors do not fade and its contours do not bend. What's more, the workers only use non-toxic glues and dyes on the toys. For packaging and promotional purposes, the company only uses recycled or recyclable materials.

Recent developments have led Plan Toys even further down the road of sustainability: the company now creates valuable composite materials from the waste products of their existing processes. Waste wood, bark and small branches unsuitable for manufacturing are used as thermal fuels to drive the factory's kilns and other processes. The scrap wood is burned in a five-megawatt biomass gasification waste-to-energy plant that the company built in 2010 next to its toy factory at a cost of more than 350 million baht. All in all, the biomass plant and solar power panels installed at the site produce enough electricity to power factory operations, which creates savings of more than 15 million baht a year on energy consumption. Each year, the company also plants about 9,000 trees that help absorb over 2,700,000kg of CO_2. The factory is also able to supply several nearby villages with power.

To counter increased competition from cheap Chinese-made toys and weaker demand from its major market, the US, the company launched a new line of toys made of "PlanWood", a non-toxic composite wood made of the sawdust that piled up in Plan's factories. The new material saved the company 32 percent on manufacturing costs, and in 2014, the sale of PlanWood toys accounted for 40 percent of their overall profits.

For this forward-looking company, sustainability principles and green manufacturing ideas inform every step of the manufacturing chain, from the early design stages to production, packaging and marketing. Plan Toys has also successfully tapped into a growing market for environmentally sound toys and built a global brand. •

tion across Thailand – inspiring them to recycle, buy local and be more responsible consumers.

In the past, companies had a tendency to think of sustainability as a luxury or as a temporary solution to deflect accusations of poor social and environmental performance. But these days, Thailand's business sector is fast embracing responsible production as a way to add value and save on costs.

One way to be a responsible manufacturer is through **ethical sourcing**. Companies that follow this principle ensure that their products are sourced, manufactured and supplied without exploiting people or the environment. Thai businesses like the Jim Thompson Silk Company, which uses locally produced silk and hires 80 percent of its staff locally in the Northeast region, have made a name for themselves by placing ethical sourcing policies at the heart of their commercial branding efforts.

Poor sourcing practices, on the other hand, often come with ramifications that hit corporations where it hurts hardest: the bottom line. In Thailand, the billion-dollar fishing industry lost considerable business from developed countries after news reports emerged exposing human trafficking and the use of slave labor in the supply chain. Major companies including Charoen Pokphand (CP) Foods, Nestle, Thai Union Group and others have been accused of sourcing from trawlers or factories where human rights abuses are said to be rife. These realizations sparked public outcry, as well as boycotts of products.

That said, the overall picture in Thailand's manufacturing sector, which generates around a third of GDP and around three-quarters of exports, is rosier.

In the face of increasing consumer demand and regulatory pressure for sustainably manufactured products – especially in

Jim Thompson Silk Company uses locally produced silk and hires 80 percent of its staff locally in Thailand's Northeast region.

Thailand's key export markets of the EU and the US, which have stringent environmental standards – Thai industry leaders and government authorities see the "greening" of domestic manufacturing as a means to safeguard the existing sector and ensure sustainable growth.

"**Green manufacturing**" has been promoted in Thailand's private sector as national policy since 2011 as part of the 11th National Economic and Social Development Plan, and the kingdom promotes green manufacturing principles under the Green Industry (GI) program. Several Thailand-based manufacturers have already taken the lead in green manufacturing. SCG and Amphol Food Processing, for example, are among growing numbers of Thai companies that have implemented programs under the **ISO 14001** standard to ensure that any environmental impacts are monitored and improved.

For example, SCG's Lampang cement plant has adopted a Semi Open Cut mining process to reduce noise and dust pollution, and minimize damage to the local mountain ecosystem. SCG has implemented a "Zero Waste to Landfill" policy to minimize its indus-

Green manufacturing

A manufacturing method that minimizes waste, pollution and energy usage; utilizes sustainable materials; and/or results in a low environmental impact.

ISO 14001

The key standard within the ISO 14000 series, a set of environmental standards released by the International Organization for Standardization. ISO 14001 defines the requirements for an adequate environmental management system (EMS) implemented by any company.

trial waste. SCG's product life-cycle assessment also helps to ensure that the company's products have minimal environmental impacts.

As of 2015, more than 11,000 factories in 30 key industries had joined the Ministry of Industry's Green Industry Program, which aims to establish Thailand as a green manufacturing hub for the ASEAN region. The program has now been integrated into the International Organization for Standardization (ISO) program known as ISO 26000. So far, Thailand and more than 60 countries have adopted the program as an internationally accepted standard for corporate social responsibility.

For the past two decades, the government has also been promoting sustainable consumption and products via various labeling and certification schemes.

Green Label Thailand, based on the trailblazing German label Blue Angel, was formally launched in 1994 by the Thailand Environment Institute (TEI) and the Thailand Industrial Standards Institute (TISI) under the Ministry of Industry. Its numerous criteria are designed to distinguish the products that cause the lowest environmental impacts in their sectors. As of June 2016, the list included some 450 products in 24 product categories from 59 companies. Those figures are sure to increase in the years to come as the government has expanded its list of product categories, including fluorescent lamps, Coolmode fabric and printer toner, as well as services such as office cleaning, oil changes and automobile repairs with more than 2,000 agencies, both at national and municipal levels, participating.

Meanwhile, other labeling schemes like the energy efficiency labels for home appli-

> "A business that operates on the principle of sustainable development incorporates long-term, strategic planning that pairs business growth with positive environmental and social continuity based on fairness and equity."

ances and Green Leaf for hotels are pushing both consumers and businesses to think beyond price tags and profits. The most advanced scheme is the Green Industry Mark, which focuses on greening the entire value chain step by step, creating networks of eco-savvy manufacturers in Thailand.

GREEN PUBLIC PROCUREMENT POWER

When it comes to purchasing power, the individual wields considerable clout, but so do those in government. The procurement orders placed by government departments, authorities and state-owned enterprises make up somewhere between 15 percent and 30 percent of

SCG Packaging was the first Thai company to be certified by the Forest Stewardship Council.

SEP IN PRODUCTION

Sa Paper Preservation House is a small manufacturer of mulberry paper products operating in northern Thailand. Having grown from a family business, it now exports 80 percent of its products. Sa Paper is a strong example of a business that adheres to SEP principles, including in terms of sustainable production. Here are some details about Sa Paper's business practices and the way these help fulfill many of the calls to action of the SDGs:

Progress with balance

Economy
- Created local jobs and increased income
- Introduced more than 1,000 new products
- Made communities more economically resilient

Society
- Improved education and skill levels
- Offered interest-free loans to employees
- Created employee savings programs
- Allowed employees with families to work from home

Environment
- Conserved the environment by recycling
- Encouraged development of plantations

Culture
- Promoted traditional paper-made products

Moderation
- Regulated the pace of expansion based on the availability of capital, labor and natural resources, as opposed to surges in market demand
- Continued to produce hand-made products despite demand increases, and occasionally turned down orders

Reasonableness
- Recycled and reused mulberry paper
- Invested in a waste water treatment system

Prudence
- Promoted diversity by having a large variety of products
- Started the business in the local market, and expanded to domestic and international markets respectively
- Paid local villagers to grow mulberry trees to ensure availability of a sufficient supply

Knowledge
- Innovated to replace chemicals with natural products
- Taught people from surrounding communities and elsewhere how to make mulberry paper products

Virtues
- Has never laid off employees in its entire history, even in times of economic crises
- Paid hill tribes (the raw material suppliers) more than the normal rates

Source: Thailand Sustainable Development Foundation, 2016

---

the national GDP in bigger countries and around half of the total GDP in some developing nations.

The public sector is an important client because it places large-volume orders at regular intervals. Moreover, public expenditures incentivize innovations and investments by creating strong demand in a specific sector. Companies see such developments as stable business opportunities and, with adequate financial support, make the required long-term investments for research and development. By the time economies of scale are achieved, the eco-friendly products (like chlorine-free paper or LED light bulbs) also become available to individual consumers and a national market emerges.

Green Public Procurement (GPP) plans also send a clear political message. By spending state funds in line with declared policy goals, governments demonstrate their commitment to promoting sustainable production and consumption.

Thailand's Pollution Control Department (PCD) has been implementing GPP policies since 2005. The country's first GPP plan (2008–2011), focusing on central government and administration bodies, is estimated to have delivered reductions of up to 25,685 tons of CO_2 with the PCD spending some 62 percent of its budget on eco-friendly products.

The 2nd GPP (2013–2016) aims to increase GPP volume, stimulate more green products, support the private sector in green production, encourage governmental units to implement GPP, improve monitoring, and promote sustainable consumption in the public and private sector as well as among the general public. If these targets are accomplished, it will go a long way toward helping Thailand achieve Goal 12.

Further Reading

- A *Review on Green Manufacturing: It's Important, Methodology and Its Application* by I.D. Paul, G.P. Bhole, J.R. Chaudhari, 2014

- *An Evaluation of Green Manufacturing Technologies Based on Research Databases* by Sung-Hoon Ahn, International Journal of Precision Engineering and Manufacturing-Green Technology, 2014

13 CLIMATE ACTION

Take urgent action to combat climate change and its impacts.

CLIMATE ACTION
One of the Most Complex Challenges Humanity Has Ever Faced

Storm clouds gather over a boat off Phuket in the Andaman Sea.

Calls to Action

- Move toward eliminating dependence on fossil fuels, promote alternative energy and encourage the participation of all sectors in reducing emissions and preserving the natural environment
- Strengthen resilience and adaptive capacity to climate-related hazards and natural disasters
- Further integrate climate change measures into national policies, strategies, budgets and planning and enhance coordination between government agencies
- Create effective financial mechanisms to support the implementation of climate change mitigation plans
- Improve education, awareness-raising and human and institutional capacity on climate change mitigation, adaptation, impact reduction and early warning
- Implement the commitment undertaken by developed-country parties to the United Nations Framework Convention on Climate Change

Decimated rice crops, Bangkok under water and large swaths of coastal tourism wiped out by storm surges: such climate change doomsday scenarios for Thailand may seem like Hollywood blockbuster fodder, if you don't believe in science. Unfortunately, study after study is proving that human activity is altering our planet and environment in unprecedented fashion. A rising concentration of greenhouse gases in the atmosphere, caused largely by global industry and agriculture, is cooking the Earth and making our oceans more acidic. The clearing of forests for farming has destroyed biodiversity and ecosystems – and further exacerbated the greenhouse gas threat – and the depletion of freshwater resources to meet the skyrocketing demand of both the public and private sector means that the element primary to all human endeavor – water – could become the subject of destabilizing competition in the future.

An alarming report from the Intergovernmental Panel on Climate Change released in 2014 warns that if the current usage of fossils fuels continues to accelerate, the average temperature of the

Earth's surface will rise by 1.4–2.6 degrees Celsius by the middle of the 21st century. Scientists predict this rise in the global average temperature will lead to mega-droughts, mega-floods, violent storms, species extinction, crop failures, a massive rise in sea levels

> **"We have reached a level of CO_2 in the atmosphere not seen for the past 3 million years."**
> -Jeffrey D. Sachs, Earth Institute director

and the stunted growth of marine life as the oceans absorb more and more CO_2. In Thailand, as everywhere, plants and animals that are unable to adapt to warmer climates may be threatened with extinction. Warming temperatures may "adversely affect rice and other crops", the report says. Meanwhile, climate change could accelerate the spread of infectious diseases like malaria or dengue fever as parasites and bacteria find a more favorable environment in which to replicate.

As the Director of the Earth Institute, Jeffrey D. Sachs, writes in *The Age of Sustainable Development:* "There has never been a global economic problem as complicated as climate change. It is simply the toughest public policy problem that humanity has ever faced. First, it is an absolutely global crisis. Climate change affects every part of the planet, and there is no escaping from its severity and threat. Humanity in the modern period has faced some pretty terrible threats, including nuclear annihilation along with mass pandemic diseases. Climate change ranks right up there on the scale of risks, especially for future generations."

The terrifying threat of climate change was made clear by World Meteorological Organization Secretary-General Michel Jarraud in September 2014, after concentrations of carbon dioxide in the atmosphere reached a record high of 142 percent more than the pre-industrial era. In an impassioned plea to the international community, he warned that we must take concerted action now or face potentially devastating consequences. "We know without any doubt that our climate is changing and our weather is becoming more extreme due to human activities such as the burning of fossil fuels," he said. "We are running out of time."

THE THREATS TO THAILAND

A regional climate projection conducted by Chulalongkorn University predicted that, as global warming increases, Thailand would have higher rainfall in the order of 10–20 percent in all regions, although the number of days with rainfall is not expected to change significantly. In addition, all regions will be warmer and the duration of the cold season will be shortened, while extreme events – especially droughts, floods, storms and landslides – will intensify.

In Chainat province, farmers and their crops suffered significantly during the 2015 drought.

Avoiding the Two-Degree Tipping Point

In the field of climate change, two degrees Celsius has become both a warning sign and a magic number, for scientists believe that if global temperatures rise any more than that above pre-industrial levels then the planet will be on the verge of a catastrophic tipping point. As far back as 1975, the Yale professor of economics, William Nordhaus, had said that a global mean temperature increase of two or three degrees would be without precedent in the "last several hundred thousand years", writing in an abstract for the paper, "Can We Control Carbon Dioxide?": "It appears that emissions of carbon dioxide particulate matter, and waste heat may, at some time in the future, lead to significant climatic modifications."

In a 1990 report by the Stockholm Environment Institute, the upper limit of two degrees was affirmed, though the report warned that even a rise of one degree "could lead to extensive ecosystem damage". When 115 world leaders and thousands of NGO staff, scientists and the media descended on Denmark for the Copenhagen Climate Change Conference in 2009, that figure was the barometer of a debate that disappointed many by not setting any binding targets. In fact, the two-degree limit did not become international climate change policy until the Cancun Agreements of 2010. For the 2015 UN Climate Change Conference in Paris, as part of the Intended Nationally Determined Contributions (INDCs), the kingdom has declared its commitment to help reduce global warming by cutting greenhouse gas emissions by 20 to 25 percent by 2030. At the conference, the member states adopted the Paris Agreement, a universal agreement whose aim is to keep a global temperature rise for this century well below two degrees Celsius and to drive efforts to limit the temperature increase even further to 1.5 degrees Celsius above pre-industrial levels. The agreement, which was ratified by Thailand as well as major world powers China and the US in 2016, "recognizes that climate change represents an urgent and potentially irreversible threat to human societies and the planet and thus requires the widest possible cooperation by all countries, and their participation in an effective and appropriate international response, with a view to accelerating the reduction of global greenhouse gas emissions."

Unfortunately, the impact of these and other changes will in all likelihood severely disrupt three of Thailand's key economic mainstays: agriculture, tourism and trade. According to a UNEP analysis, Thailand is one of the countries most at risk when it comes to climate change's impacts on its agricultural production, with an estimated loss of between 15 and 55 percent by 2080 possible. As usual in Thailand, floods and droughts of increasing severity will be the likely culprits. Between 1989 and 2012, there were 227 floods in Thailand. These hundreds of floods claimed more than 4,000 lives and destroyed more than a million homes, according to the 2012 Statistics on Disaster by the Department of Disaster Prevention and Mitigation (DDPM) under the Ministry of Interior. Droughts have also plagued the country's agrarian heartland and the arid Northeast, costing the country some 15 billion baht and affecting some 10.5 million hectares of farmland. Given that the country is the world's largest exporter of rice and the agricultural sector employs nearly 40 percent of the population, drastic disruptions in agricultural production would likely have severe local and regional repercussions, hitting the most vulnerable farmers such as small-scale farmers the hardest, and potentially leading to higher food prices and economic hardships that could cause civil unrest.

Thailand's oceans, seas and coastal areas

Thailand Has Pledged to Cut Greenhouse Gas Emissions BY 20 TO 25 PERCENT BY 2030

also face multiple threats. Through overfishing and overharvesting of marine life, human activity is already putting tremendous pressures on marine ecosystems. Climate change will make matters worse. Violent storms could disrupt the kingdom's tourism industry, which is centered around southern Thailand's islands and beaches. The tourism industry as a whole contributed nearly six percent directly to GDP as of 2015 and employed nearly 11 percent of the total workforce. Even if such storms do not occur, the rising concentration of CO_2 will be absorbed by the oceans, increasing its acidity, which will further kill off coral reefs and stunt the growth of shellfish crabs, lobsters and other marine life. Such changes could further exacerbate the struggles of Thailand's traditional fishing industry.

Bangkok, a crossroads of trade for the country and region but one of the world's most low-lying cities, is also vulnerable. In the 2013 Climate Change Vulnerability Index prepared by Maplecroft — a risk analysis firm for international corporations and many UN organizations — Bangkok placed third after Dhaka and Manila in a survey of major cities sure to be affected by climate change. (Thailand ranked 45th most at risk out of all the countries in the world.) For one thing, Thailand's capital sits in the midst of a gigantic flood plain and river delta. For another, it sinks, according to one analysis, between 20 and 28 millimeters per year, a rate that has doubled in the last decade, according to an estimate by a Chulalongkorn University professor. Surface erosion and excessive groundwater pumping, mostly by industries, housing estates and small businesses, are the main culprits. With ground water access entirely unregulated by the government and urban sprawl only growing, Bangkok looks set to sink further in the next decades, with its longest street of Sukhumvit Road potentially under threat of crumbling to erosion.

CLIMATE ACTIONS

Because climate change is a global crisis without boundaries, mobilizing people in Thailand and around the world to take corrective action is immensely challenging. As climate change is a slow-moving crisis, even convincing people that it is a challenge to address at all can be difficult. Doubts about the veracity of the science, short-term considerations and short-term profits have often trumped the long-term visions and implementation that climate change activists have urged. Politicians, CEOs and leaders tend to focus on the immediate demands of their constituents or stakeholders, or on solving current crises. Those who are likely to be the victims of climate change's more dire consequences don't have a voice because they have not even been born yet. So actions are easily postponed. Moreover, as the biggest causes of greenhouse gas emissions originate from those sectors that are at the core of the modern economy (such as energy and agriculture), it can be difficult to convince members of

Mae Moh coal-fired power plant in Lampang province. Despite its numerous negative impacts on the environment, Thailand continues to pursue coal projects.

The 2011 floods impacted communities across the country.

government and the private sector to take drastic action. Companies in these sectors typically act as powerful lobbyists against new regulations, although there are signs that the private sector is beginning to consider the risks posed by climate change and rally to the cause of sustainable development. In short, there is a lot of resistance to acting on the threat of climate change.

Moreover, solutions are available but they will be complex. Categorized broadly, the collective action of individuals, businesses and governments will be the key to reversing current trends. Fighting climate change may seem beyond the scope of individual action but because everyone is a contributor, everyone can help. Through awareness and moderation of our own personal consumption or the local actions we take – whether in our home, business or community – we can make a difference. In fact, consumers today have never had so much power at their disposal. Never has voting with your wallet and conscience been as popular, or as effective, with brands both big and small. Every consumer choice we make – from what type of disinfectant we use to what light bulb we purchase to where we dine out, has an effect on both the environment and the bottom line of companies that market these products or services.

Even in our daily lives, the decision whether or not to take public transport to work will save fossil fuels or add more greenhouse gas emissions to the mix. Refusing to accept that plastic bag at the convenience store or choosing a more eco-friendly holiday rather than a luxurious shopping spree in a foreign country – all these little decisions inform the bigger picture. Consumer power is pushing the private sector toward practices and products that are environmentally friendly and do not exploit human capital, or contain harmful chemicals, for example. The wealthy have a

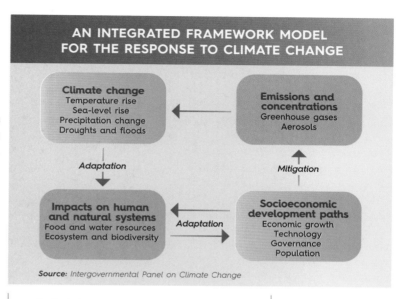

Source: *Intergovernmental Panel on Climate Change*

particularly important role to play if they are willing to change their mindset and behavior. An Oxfam report released in 2015 stated that the world's richest ten percent produce half of all carbon emissions.

In Thailand, the Sufficiency Economy Philosophy (SEP) is one effort that has helped create such mindfulness and awareness among individuals by emphasizing the benefits of moderation, reasonableness and prudence. The Thai government has used SEP as a framework for its Thailand Climate Change Master Plan 2012–2050. This plan provides a blueprint for, and guidelines on, climate change preparedness, adaptation, and other appropriate and efficient applications in the name of creating a sufficiency economy and low carbon society.

The latter goal is an increasingly urgent matter. Greenhouse gas emissions reached a staggering 278 million tons in 2010, according to an article called "World Carbon Emissions: The League Table of Every Country", published in *The Guardian*. The most practical way to lower emissions of CO_2 and other greenhouse gases is to support the production of renewable energy sources, promote energy

2014 and 2015 Each Set the Record FOR HOTTEST CALENDAR YEAR SINCE WE BEGAN MEASURING SURFACE TEMPERATURES OVER 150 YEARS AGO

conservation through the reduction of power usage and lay the foundations for a more environmentally friendly society.

Thailand, however, remains heavily dependent on fossil fuels, with 92 percent of electricity production depending on them, and it ranks as the 23rd heaviest emitter. The government refuses to rule out coal as a source of energy in spite of its obvious deleterious effects on global warming, and has recently gone ahead with plans to construct six coal-powered plants. At the same time, there are encouraging trends. The Thailand Climate Change Master Plan is committed to the polluter pays principle, and since 2007, the government has been providing an incentive in the form of feed-in tariffs (or FIT) to boost

> "What Thailand needs are sustainable solutions on multiple scales, with better and more integrated water resource management, regional land zoning and public participation at a community level to address disaster risk reduction."

private investment and encourage public participation in renewable energy, especially solar. Today, 33 percent of total energy production is from renewable resources and Thailand ranks 12th in renewable energy production (out of 141 countries), according to the International Energy Agency's Energy Atlas. Thailand's Energy Efficiency Development Plan (2015–2036) aims at reducing energy intensity by 30 percent (as compared to 2010), while the Alternative Energy Development Plan (2015–2036) calls for boosting the use of alternative sources to 30 percent.

On the global level, in 1999 Thailand ratified the Kyoto Protocol (1997), which

Climate Change and Rice Production

Rice is Thailand's staple crop, key to the diets and incomes of tens of millions, and the country has consistently been the world's top exporter. In 2015, the kingdom yielded over 26 million tons and exported 9.8 million tons. Unfortunately, rice is also a contributor to climate change, with the heavy use of chemical fertilizers many farmers employ partly to blame. A move toward organic production of rice would seem to be the ideal solution. However, in an illustration of the complex challenges posed by climate change, even organic rice production would contribute to global warming. This is because, while global warming is commonly associated with carbon dioxide emissions, there are other greenhouse gases that contribute to global warming such as methane, which accounts for 14 percent of emissions worldwide and traps 72 times more heat than carbon dioxide does over a period of 20 years. And half of all methane released in Thailand derives from the cultivation of rice, be it organic or not. That's because when organic matter ferments in flooded paddies, methane is released. So as Corrine Kisner's article "Climate Change in Thailand: Impacts and Adaptation Strategies" points out, "organic fertilizer alone doesn't provide the climate solution for rice (although it greatly improves farmer health and soil fertility)…"

So what to do? As with many climate change issues, the solutions tend to require more complex management. One option is for farmers to transform the biomass that results from rice cultivation, such as rice husks, as a source of biofuel, or to produce heat or electricity from the biomatter. Indeed, biomass energy is a burgeoning industry in Thailand and the government is supportive. Yet even biomass, if handled carelessly, can have harmful effects that threaten biodiversity and bode ill for public health. Another strategy Kisner points to in her article is to occasionally drain the paddies, thereby eliminating the bacteria that thrive in the oxygen-free setting and that produce methane by decomposing manure or other organic matter. The government has also been supportive of this idea. However, implementing such practices across Thailand's hundreds of thousands of small-scale rice-farming plots is not easy. As the issue of the cultivation of rice in Thailand demonstrates, mitigating climate change has no simple solutions but there are solutions possible. What is often required is awareness, innovation and proper management.

extended the historic 1992 United Nations Framework Convention on Climate Change (UNFCCC). It also ratified the Paris Agreement in 2016, pledging to help reduce global warming by cutting greenhouse gas emissions by 20 to 25 percent by 2030. The fact that Thailand still relies heavily on fossil fuels and may face opposition from a variety of powerful interests that profit from this industry means that fulfilling the Paris Agreement will be a great political challenge for the kingdom, requiring cooperation from the private sector and state-owned enterprises.

More than pledges, what Thailand needs are sustainable solutions on multiple scales, with better and more integrated water resource management, regional land zoning and public participation at a community level to address disaster risk reduction. This is not what has been happening. Successive governments have not changed their flood-prevention strategies, which remain heavy on structural measures. Government plans to cope with droughts must be moved from the drawing board of hypotheses to workable realities, while farmers must diversify crops to be less reliant on high levels of water consumption.

There are other actions that need to be taken by the public sector. Among them, the preservation of mangrove forests, which serve as bulwarks against surging waters and coastal erosion; better enforcement of policies to protect forests and swamps in order to increase their capacity to retain or drain floodwaters; the implementation of more early warning systems such as the one set up in the wake of the 2004 tsunami; and further encouragement of public participation as well as the dissemination of relevant information by state media to create further awareness.

Given the lack of political continuity in Thailand, relying on the politicians to take climate change seriously may be risky. Thus it is often emphasized that the powerful private sector take the lead. According to the Thailand Climate Public Expenditure and Institutional Review that examined Thailand's policies and budget allocations for climate action, "mitigation actions will depend critically on private sector engagement, particularly in the clean energy sector…" Fortunately for Thailand, its resilient private sector, no stranger to mitigating the consequences of unpredictable events, is addressing the challenge of sustainable development and climate change with much more sincerity in recent years.

Part of the impetus is the increasing adoption of SEP principles in the private sector and the growing desire to be recognized as a company dedicated to sustainability. Thailand has 15 companies currently listed on the Dow Jones Sustainability Index (DJSI), and many are involved in either the energy, agriculture or marine products sectors such as Banpu, Charoen Pokphand Foods (CPF), PTT, PTT EP, SCG, Thai Oil and Thai Union Frozen (TUF). These companies are among the most profitable and largest not only in Thailand but in Southeast Asia. Companies like TUF, which was blasted for unethical labor sourcing along its supply chain, have made very public commitments to eradicate the practice. Thai Oil was named the Energy Group Industry Leader by the DJSI. Meanwhile, SCG has been acclaimed for offsetting the environmental impacts of its cement and paper plants by creating waste-heat recovery facilities, which produce energy from heat emissions, and building check dams in forests to protect watersheds and preserve ecosystems. The company is also renowned for innovating eco-friendly products. Throughout Thailand, companies are increasingly re-examining their operations and practices to take into account their ultimate external impacts.

Further Reading

• *The Age of Sustainable Development* by Jeffrey D. Sachs, 2015

• *Thailand Climate Public Expenditure and Institutional Review*, 2012

• *Thailand Climate Change Master Plan 2012–2050*

• *Six Degrees: Our Future on a Hotter Planet* by Mark Lynas, 2007

• *This Changes Everything* by Naomi Klein, 2014

14 LIFE BELOW WATER

Conserve and sustainably use the oceans, seas and marine resources for sustainable development.

LIFE BELOW WATER
Safeguarding the Oceans for Future Generations

Calls to Action

- Prevent and reduce marine pollution from coastal land-based sources and man-made activities including untreated industrial and agriculture waste, consumer trash, oil spills and sewage

- Improve management and protection of marine and coastal ecosystems

- Curb land reclamation and encroachment on mangroves; mitigate the effects of tourism encroachment and wave-induced coastal erosion

- Minimize and address the impacts of ocean acidification, including through enhanced scientific cooperation

- Empower small-scale fishers by promoting access to marine resources and markets, and by teaching sustainable fishing practices

- Regulate fish harvesting and end overfishing; curb illegal and unregulated fishing and fisheries; and stop destructive fishing practices

- Increase scientific knowledge, develop greater research capacity and improve international cooperation

Koh Panyee fishing village in Phang Nga province.

Oceans are the lifeblood of our planet. They influence our weather systems, help regulate climate change, provide much of our food and water, clean our air and fuel our economies. As vital as oceans are to our continued ability to inhabit this planet, you'd think we would have taken better care of them. Unfortunately, greenhouse gas emissions, pollution and overfishing have put these vast resources at grave risk across the globe.

According to Earth Institute Director Jeffrey D. Sachs, "If we do not take care and face up to these multiple assaults, we will face growing crises in the not-so-distant future." The good news is that managing these resources responsibly is, in theory, within our grasp. Put simply, we know what needs to be done. But because the oceans are so interconnected, their sustainable management requires a sincere "global partnership", which has yet to materialize on a meaningful level. What's needed is for all relevant stakeholders — including nation states, seafood companies, extractive industries, environmental groups, coastal communities and individuals like yourself — to work cogently to manage marine and coastal resources in a way that balances economic needs and demand with environmental protection. Across the globe – and this holds true in Thailand as well – proper cooperation on the sustainable management of marine and coastal resources is still somewhat lacking. Put simply, we can do better.

> "About 8 million tons of plastic are dumped into the world's oceans every year, the majority of which comes from just five countries: China, Indonesia, the Philippines, Thailand and Vietnam."

The story of Thailand's oceans, seas and 3,200 kilometers of coastline captures the tensions of sustainable development in a clamshell. On the one hand, the rich natural resources of coastal areas have been instrumental in driving growth, especially since the 1970s. Oil and natural gas reserves, the enormous tourism industry and one of the world's biggest seafood industries are now the basis of millions of jobs and are all major contributors to the modern-day Thai economy. In their wake, however, these developments have left a significant level of environmental degradation, which Thailand now finds itself struggling to address adequately. According to the SDG Index, Thailand's "Ocean Health Index" scores, which cover the health of its waters, biodiversity and fisheries, are all firmly middle-of-the-pack when compared to the rest of the world. In addition, 43 percent of fish stocks have collapsed or are over-exploited. Overall, Thailand does score higher than regional averages.

Destructive fishing methods such as bottom trawling have destroyed many coral reefs – the world's second-most productive ecosystems after rainforests – and substantially reduced coral cover. Coral bleaching and the dip in bigger marine creatures like mantas and whale sharks have also threatened the kingdom's reputation as one of the world's best dive spots.

Meanwhile, capitalism and consumerism have spurred increased use of plastic and plastic-intensive goods. Unfortunately, after such items have served their purpose and are discarded they are increasingly finding their

MARINE RESOURCES

Total Coastline: 3,200 km

Provinces with Coastal Areas: 23

Total Marine Area: 316,118 sq km

Islands: 936

The Andaman Sea Coast is characterized by deep oceanic waters and a narrow, rocky and coral-reef-associated continental shelf, with a thick mangrove belt protecting the coastline.

The Gulf of Thailand has a shallower profile with a combination of mangrove forests, mudflats and sandy beaches. It is highly productive for fishing due to its shallow depth and high influx of nutrients and freshwater from regional rivers.

Source: Department of Marine and Coastal Resources; Greenpeace

way into our oceans. Thailand ranks sixth globally (out of 192 coastal countries) in the mismanagement of plastic waste, which has resulted in 1.03 million tons of plastic debris sinking to the bottom of the ocean along the country's coastlines, according to a study published in *Science Magazine* in 2015. Another study, conducted by the Ocean Conservancy and McKinsey Center for Business and Environment, puts the kingdom in the top five of worst offenders, along with China, Indonesia, the Philippines and Vietnam, who, together, account for 60 percent of plastic pollution in the oceans. Unless steps are taken to manage this waste properly, in ten years' time the oceans could contain one ton of plastic for every three tons of fish.

Compounding the kingdom's coastal and marine management woes is the high level of industrial activity along the coasts. In previous decades, discoveries of natural gas and oil in the Gulf of Thailand fueled the industrialization of the country's Eastern Seaboard and the growth of the nation as a whole, with the state-run Map Ta Phut Industrial Estate and the deep-water port in Rayong province built to host petrochemical companies and other heavy industries. Despite the existence of many laws and regulations, the wastewater released by industrial plants in these areas and elsewhere often exceeds the permissible limit. Unfortunately, since 2004, there have also been more than a dozen oil spills and leakages. As with so many other environmental issues either on land or at sea, the long-term effects of these pollutants on their respective environs and the creatures that live there are as yet unknown.

On a micro-level, Thailand has numerous success stories (discussed later in this chapter) that show how the kingdom is making efforts to right past wrongs and work toward more sustainable management of ocean resources. But in practical terms, while these projects may be undertaken with altruistic intentions, they lack the capacity required to make a game-changing impact. In short, much larger-scale efforts are required to stem the tide of environmental degradation on and off Thailand's coasts.

Coral cover
The amount of stony coral that exists on a reef. It's both the reef's main building block and an important habitat for many marine creatures.

Ocean acidification
At least one-third of the carbon dioxide emitted from cars, factories and buildings is absorbed by the oceans, which raises their acidity while lowering their ability to safely absorb more carbon. This process of acidification has had a profoundly negative effect on marine life like corals.

Improving waste management in China, Indonesia, the Philippines, Thailand and Vietnam can **reduce global plastic pollution in oceans by**

45% by 2025

Raising collection rates across China, Indonesia, the Philippines, Thailand and Vietnam to 78% would lead to a

23% reduction in plastic leakage into the oceans.

Improved collection infrastructure and plugging collection gaps can **reduce annual leakage by nearly**

50% by 2020

Source: Ocean Conservancy

ROCKING THE BOAT TO PROTECT FISH STOCKS

Thailand's coasts are home to 23 sea-straddling provinces, where both wallets and stomachs are inextricably linked to marine fisheries, coastal aquaculture, agriculture and eco-tourism. It should come as no surprise then that this dependence has led to more coastal communities taking on stewardship roles in the management of marine ecosystems. And as awareness grows about the negative impacts heavy industry can have on the environment, an increasing number of coastal residents are deciding to fight for more sustainable development practices, namely by mobilizing their peers to call for changes to government policies that would allow them to have a say in the state's management of the marine and coastal resources.

For example, the contentious Southern Seaboard Development Project (SSDP), the fate of which is still in limbo, has run up against significant opposition from activists and local communities concerned that the resource-rich waters off the coasts of Songkhla and Satun provinces could be damaged by a new wave of development.

Part of the SSDP entails building a 142-kilometer railway linking the Andaman Sea with the Gulf of Thailand to facilitate trade with neighboring countries and the rest of Asia. This will necessitate constructing a deep-sea port on each coast, one in Songkhla province and one in Satun. As alternatives to Map Ta Phut, these could provide much-needed sources of employment and revenue. But locals worry that the mega-projects may threaten two of the region's biggest earners – fishing and tourism – also displacing communities that rely on these sectors for their livelihoods. In addition, a portion of the construction in Satun province would take place in Mu Ko Phetra National Park, raising further questions about the damage that could be done in this protected area and beyond.

National Reform Council member and marine biologist Thon Thamrongnawasawat cautioned in April 2015 that if, as planned, the Satun port is constructed within the park, then marine ecosystems as far away as Phang Nga Bay and Koh Lipe island could be devastated, meaning the area's "rich biodiversity will vanish forever".

While the government has assured that the port will be "green" and has also enacted a new fisheries act – the Royal Ordinance on Fisheries B.E. 2558 (2015) – to protect marine resources, locals still fear that if the project goes ahead it will open the floodgates to more such schemes.

Up and down the coasts, development projects have already taken a heavy toll on traditional fishing, thanks to an abundance of wastewater discharges polluting the seas. These small-scale fishing communities have also lost business and access to fish stocks due to the prevalence of bigger trawlers that their low-tech vessels can't compete with, according to a 2013 report by Greenpeace.

Wholesale fish market in Samut Sakhon province. Across the globe, fish stocks are being depleted.

> "As our oceans empty in what is the biggest mass extinction of species since the age of the dinosaur, we should ask ourselves: Do we know where that seafood on our plates came from, who caught it, which middleman bought it and who packed it under what sort of conditions?"

Globally speaking, the impacts of commercial fishing are even more dire. We've seen a 26 percent rise in ocean acidification since the start of the industrial revolution. Fish stocks have been decimated, and marine biologists believe that 90 percent of bigger species like tuna have already been fished out. Some marine scientists forecast that if there is not a significant re-evaluation and slowdown of overfishing soon, the planet's fish stocks will be mostly gone by 2048. The direct ramifications for humans are quite daunting when you consider that more than three billion people depend on the oceans for their livelihoods and primary source of protein.

Achieving sustainability in the fishing sector is certainly crucial for Asia, where around two-thirds of the world's seafood is consumed and where fishing fleets, including those in Thai waters, are notorious for destructive techniques like bottom trawling. The purse seine nets used by some industrial trawlers scoop up everything in their path, including coral reefs, eggs and juvenile fish, so that 70 percent of each haul is referred to as "bycatch" or "trash fish" and ends up as animal feed or sold under generic names like "sea bass".

> "As much as 40 percent of the world's oceans are heavily affected by human activities, including pollution, depleted fisheries and loss of coastal habitats."

Why Oceans and Seas Matter to Sustainable Development

SOAKING UP CARBON Oceans and seas are the biggest "carbon sinks" on the planet, meaning they soak up a quarter of all the carbon dioxide emitted by cars, factories and cities.

NET PROFITS Fisheries and aquaculture are two of the country's economic lifelines with Thailand ranking number three out of the world's top seafood producers.

SHELTER FROM STORMS Seagrass beds, coral reefs and mangroves are sanctuaries for marine animals. They enhance nutrient circulation in the ecosystem, filter wastes and protect shorelines from strong water currents.

NATURAL ASSETS The Gulf of Thailand is rich in crude oil, natural gas and valuable minerals such as tin.

TOURISM HOTSPOTS Many tourists visiting Thailand make a beeline for the country's fabled islands and beaches. According to government statistics, Thailand's international tourism revenue reached more than US$60 billion in 2015, while tourism arrivals reached 29.88 million that year.

FINITE FISH More than three billion people depend on the oceans for their primary source of protein.

CORAL COLLAPSE Coral reefs, also known as the "rainforests of the sea", are some of the most biodiverse and productive ecosystems on earth. But an estimated 75 percent of coral reefs are threatened by factors such as destructive fishing methods and rising sea temperatures.

The Pak Phanang River Basin Royal Development Project Success Story

The Uthokawiphatprasit Watergate at Pak Phanang River Basin.

Achieving the sustainable management of water sometimes requires leaders to be creative in order to find common ground between parties with polarizing views on how a particular resource should be used. The Pak Phanang River Basin Royal Development Project is a classic example of one such success story.

Situated in a shallow bay below sea level, the Pak Phanang River Basin covers a total area of more than 300,000 hectares and is home to around 600,000 people. The basin, which covers parts of four different provinces, is fed by freshwater, brackish water and saltwater. Once a fertile plain known as the rice bowl of southern Thailand, the basin began to deteriorate in the 1980s due to urban encroachment and the destruction of the watershed, which triggered heavy floods during the monsoon season and disastrous droughts during the dry season.

Many rice farmers converted their rice fields into shrimp aquaculture, an export-oriented business that has boomed in Thailand in recent decades as demand for shrimp soared around the world. However, intensive shrimp aquaculture typically relies on antibiotics and chemicals to increase yields and leads to saltwater seepage, polluting the groundwater and ecosystem, rendering it unsustainable for farming in the future. Wastewater from shrimp ponds was not treated properly in the Pak Phanang River Basin and was released into the river, damaging agricultural land, especially paddy fields, and forcing some rice farmers to leave their land. Diseases among shrimp populations, which are common due to the monocropping nature, can cause the spread of pathogens in wetlands.

All these problems ultimately boiled over into a prolonged local conflict, pitting the freshwater farmers against the shrimp farmers. Into the fray stepped King Bhumibol Adulyadej. Following his advice, the Uthokawiphatprasit Watergate was constructed from 1996 to 1999 about three kilometers from the Pak Phanang district. This watergate would address the problems of drought, flooding, salinity and a lack of freshwater.

Its main functions were twofold: preventing saltwater from seeping into the river and contaminating farmland, as well as storing freshwater to use for agricultural purposes and household consumption. The project also included the construction of a long embankment to separate freshwater and saltwater, and an irrigation system to tackle flooding and water pollution by channeling water to a special holding tank in the basin.

Zoning was another integral part of the plan. This meant that farmers residing in the different zoning areas had to consider changing their crops based on the type of water available in their localities. For example, shrimp farmers converted their saltwater farms into freshwater agricultural plots. Zoning has helped end prolonged local conflicts. In addition, the watergate has also mitigated flooding in the area, while the saltwater irrigation has helped increase the productivity of shrimp aquaculture.

The development project has not only restored the livelihoods of local people but also boosted their revenues. Since 2004 the average income of the locals has increased by 28 percent. Overall, the Pak Phanang River Basin Royal Development Project's greatest strength lies in its multifaceted approach that has boosted local standards of living just as it's been a boon for the environment. •

Coral colony on a reef at Koh Chang.

Fortunately, efforts are underway in Thailand to produce higher quality fish in sustainable ways. The Earth Net Foundation's Small Scale Fishers and Organic Fisheries Products Project was started with funding from the EU to address this shortage. The main goals are to support small-scale fishermen who are doing conservation work, and to help them obtain better prices for their products – for example, getting 70 percent of the value instead of the 20 to 30 percent they are usually offered by middlemen. Another organization, the Organic Agriculture Certification Thailand, is promoting safe, eco-friendly and socially responsible fish catches through a certification program with several criteria for sustainability.

While the motivation behind such initiatives is commendable, they would have to be scaled up significantly to make an impact that would lead to a rebound in fish stock numbers.

MARINE CONSERVATION TAKES TEAMWORK

Until the 1980s, marine preservation was not on the radar for Thailand's politicos and bureaucrats. However, a growing, if diffuse, awareness of the myriad benefits (fiscal, environmental and social) that such spawning grounds provide for a range of development sectors was acknowledged with the designation of marine and coastal protected areas. Today, it's much more widely understood that these areas, such as marine national parks and restricted areas for projects like turtle breeding, are essential for maintaining the capacity of critical ecosystems to support sustainable development.

One result is that combating threats in these areas has united an array of talents, from economic planners and policymakers to members of the private sector, who have teamed up with academics, activists and local residents. Such collaborations have made it possible for a number of projects to proceed. In Trat province, the Ban Pred Nai Community Forestry Group has banded together to preserve the area's mangrove forests. Renowned for their abilities to stop soil erosion and protect villages from storms, mangroves are also known to protect juvenile marine creatures that hide in their tangled roots. By protecting these forests, the villagers have also protected their own livelihoods as fishermen and preserved their local cultures.

Many seaside communities like these are dependent on fishing for both cash and calories. In the southern province of Chumphon, a crab bank has paid dividends for locals who are now well aware of the Ban Pred Nai Community Forestry Group's pragmatism as reflected in the saying: "Stop catching a hundred – wait for a million." This could well be a rallying cry for sustainable development among such rural communities in Thailand, but it would fall on deaf ears if it were not for financial incentives. Both of these projects have used a system of financial bonuses and penalties to make sure villagers cooperate. Both have also set up community savings funds to help farmers in distressful situations, but only if they play by the rules.

As far back as the 1980s, King Bhumibol Adulyadej recognized the need for better marine resource management. Concerned about the deterioration of mangroves and the correspondingly negative impact on livelihoods, in 1981 the king established the Kung Krabaen Bay Royal Development Study Center, with the center and related projects covering 13,090 hectares in Chanthaburi province. Focusing on aquaculture, mangrove conservation and improving the occupational skills of local people, the center has had notable success in carrying out mangrove restoration initiatives, promoting eco-tourism

"The escalating destruction and degradation of mangroves, driven by land conversion for aquaculture and agriculture, coastal development and pollution, is occurring at an alarming rate, with over a quarter of the earth's original mangrove cover now lost."

-UN Under-Secretary-General and former UNEP Executive Director Achim Steiner

Thailand Is the Third Biggest SEAFOOD PRODUCER IN THE WORLD

Transplanting Coral Reefs

Plastic PVC pipes are used as a base to help coral regenerate.

To be an innovator in sustainable development does not always require inventing some fabulous new high-tech device or pioneering a new school of philosophy. Sometimes all it takes is putting an ingenious spin on an already existing product, like the PVC pipe, ubiquitous in Thailand, combined with a well-known technique used in dentistry to fill root canals.

This was the double-edged thrust and modus operandi of Prasan Sangpaitoon's work to revitalize the coral reef colonies along the east coast of Thailand, which had been battered by fishing, tourism and storms that bleached or ruined the reefs. Prasan, a lecturer at the Rambhai Barni Rajabhat University in Chanthaburi province, used PVC pipes as breeding grounds for reef fragments to grow in the seabed that is their natural home. "The PVC pipes we chose are the same as those used for supplying drinking water supplies. To me, if they're safe for humans they should not poison the reefs," he reasoned.

The process is simple. Prasan attaches PVC tubes to a small window-sized frame, then puts a coral fragment into each hole in the frame and attaches it with a screw. After that he puts the frame in the seabed. Unlike metal bars, which could turn rusty and kill the coral, PVC pipes are durable and non-corrosive. They're also cheap, with each PVC frame costing only about 500 baht (US$14.25). Still, it's a slow process: Corals only grow by around 1.5 to two inches a year. That's a large part of the reason why they are so grievously endangered. A World Resources Institute Report from 2011 claimed that 75 percent of the world's reefs are under threat: a worrying rise of 30 percent over the previous decade.

Knowing the importance of his task supplied much of the motivation for the professor. For his first forays into coral transplantation he chose the Samaesarn beach area in Chonburi province. After observing excellent results there, he expanded his venture to other nearby islands, such as Koh Kham, Koh Samet, Koh Wai and Koh Talu. Choosing the right coral for these underwater terrains (in this case the *Acropora* species) was decisive. Coral reefs serve as shelters and sources of food for many different marine creatures. Once these degraded areas were restored to a semblance of their former glory, the fish and other creatures returned en masse: a resurgence that also helped to shore up local fisheries. To continue and expand his work, Prasan has established a small foundation to run this successful undertaking that has now transplanted some 40,000 corals over 20 years.

Although some experts have questioned the approach of adding unnatural elements like PVC pipes or other such structures to nature, they have had to concede that this method has brought about a much higher survival rate of coral.

Prasan's project is just one of several partnerships between academics, private organizations and local communities to restore degraded coral reefs using different methods, such as sexual reproduction in an artificial environment, sinking artificial structures or reattaching coral fragments to natural substrates.

For example, the Crown Property Bureau Foundation has, since 2012, put 30 million baht toward planting 5,464 artificial reef structures off the coasts of Chanthaburi, Chumphon, Krabi, Prachuap Khiri Khan, Phang Nga and Songkhla provinces. In addition to increasing habitats for marine life, these initiatives have had the knock on effect of increasing the income of local fishermen, improving their quality of life, and prompting them to adopt more sustainable practices.

and increasing the output of local fisheries and shrimp farms.

Meanwhile, in 2013 the Reef Biology Research Group within Chulalongkorn University's Faculty of Science announced that it had successfully completed its tests to breed and release warm-water corals – the first organization to do so in Thailand. The experiment, backed by Princess Sirindhorn's Plant Genetic Conservation Project and the Royal Thai Navy, was hailed by marine scientists in the kingdom as a major breakthrough. According to the research team, the survival rate of corals transplanted with their method was 40 to 50 percent (within in the first two years), whereas the survival rate in nature is only about 0.01 percent.

Certainly it was a timely discovery. An estimated 75 percent of the world's coral reefs are now threatened. Mangrove forests, currently being depleted at a higher rate than any other type of forest, are not doing much better, though they are making something of a comeback in certain areas of the kingdom where conservation projects and protective measures hold sway.

A 2014 report from Kasetsart University said that there are some 244,000 hectares of mangroves in the country. In parts of Phetchaburi province, where the royally initiated Laem Phak Bia Environmental Research and Development Project began in 1990, they are making a slow comeback. The report pointed out that in this cluster of four villages mangroves are growing at a rate of 3.7 hectares per year. Considering the benefits that these ecosystems provide, from purifying the air and storing carbon, to serving as nurseries for juvenile fish and havens for bird-watchers and eco-tourists, they are extremely valuable. Each hectare, the report estimates, is worth around 424,000 baht (about US$12,200) per year.

The fact that many such efforts have spread to other communities and inspired many younger people is proof positive that these disparate movements, which are united by a unanimity of purpose – to preserve the coasts and seas for their descendants – will have staying power. The involvement of local and national academics and research institutions, which have lent their support through in-depth research, analyses and practical solutions, has helped strengthen the community agendas. Finally, local, national and international non-governmental organizations have also facilitated the networking and mobilization efforts and helped link the grassroots efforts to larger calls for policy changes. These combined efforts have helped fill gaps in coastal communities, which otherwise would lack science-based knowledge and modern management skills.

Far into the future, the sustainable management and protection of our oceans and marine resources will require similar across-the-board cooperation. For Thailand, following the Sufficiency Economy Philosophy – which promotes a sensitive, localized approach to the management of resources as well as reasonable and moderate consumption patterns, prudent decision-making and inclusive dialogue – could help check the over-exploitation of the country's oceans and seas. Undoubtedly, inclusive dialogue that focuses on the voices of local communities as well as the engagement of international experts will be crucial in tackling threats like marine pollution, ocean acidification, coastal land degradation and detrimental fishing practices – and the time to act is now.

Further Reading
• *The Ocean of Life* by Callum Roberts, Penguin Books, 2012

• *Stage of World Fisheries and Aquaculture: Opportunities and Challenges* by the Food and Agriculture Organization of the United Nations, 2014

• *Sustainable Intensification of Aquaculture in the Asia-Pacific Region* by the Food and Agriculture Organization of the United Nations, 2016

15 LIFE ON LAND

Protect, restore and promote sustainable use of terrestrial ecosystems, sustainably manage forests, combat desertification, and halt and reverse land degradation and halt biodiversity loss.

LIFE ON LAND

Recognizing the Importance of Soil, Forest and Biodiversity Conservation

Forests in Thailand have fallen victim to land cultivation practices, logging and fires set by poachers.

Calls to Action

- Stop deforestation and scale up efforts to restore degraded forests
- Mobilize the necessary resources to finance sustainable forest management, protect biodiversity and conserve ecosystems
- Curb monocropping and over-use of chemical fertilizers and pesticides
- Combat desertification, soil degradation and erosion
- Take urgent action to halt the loss of biodiversity; crack down on poaching and the illegal wildlife trade
- Improve forest and biodiversity protection and monitoring mechanisms
- Enforce environment and habitat protection laws

Today our planet is experiencing unprecedented land degradation and loss of biodiversity. Across the globe we're seeing the loss of arable land at 30 to 35 times the historical rate, while the persistent degradation of drylands has led to the **desertification** of 3.6 billion hectares. Additionally, 13 million hectares of forests are being lost every year and of the 8,300 known animal breeds, eight percent are extinct and 22 percent are at risk of extinction.

These figures are daunting given our reliance on fertile soil for agriculture and the fact that plant life provides 80 percent of the human diet. Forests, which cover 30 percent of the Earth's surface, also provide vital habitats for millions of species, act as key watersheds and help to mitigate the effects of climate change.

Moving forward, it's imperative that we take immediate and effective action to both rehabilitate ecosystems and ensure the long-term sustainable use of our forest and land resources. One of the key challenges ahead will be striking a balance between satisfying humankind's need for food, energy, water, minerals, medicines and raw materials without undermining biodiversity or further degrading sensitive ecosystems.

Thailand, like many other countries, is facing a number of complex environmental challenges and slowly coming to grips with the repercussions of its decades-long push to industrialize. As such, the kingdom's agricultural, industrial and aquacultural output have risen considerably but the trade-off is that its forests, rivers and soil are less healthy today, affecting their ability to host the country's remarkably rich biodiversity. The good news is that along with this adversity has come a greater awareness of these pressing issues and recognition that the country needs to take proactive measures to conserve its ecosystems.

Thailand boasts an array of natural attractions and refuges that nurture biodiversity. In total, the country has 128 national parks, 60 wildlife sanctuaries and 60 non-hunting areas covering 20 percent of the country's total landmass. Reforestation projects are beginning to reverse the drastic loss of forest cover caused by over-logging and over-farming in the last century. And King Bhumibol Adulyadej made improving water management and the quality of soil his own life's work.

MORE THAN JUST DIRT

Some people might just think of soil as "dirt", but it's much more complicated than that. Soil is a living entity – a combination of organisms, minerals, liquids and gases – that nurses plants and nourishes crops. Fertile soil is the lifeblood of agriculture, and therefore, ensuring the health of Thailand's soil is vital to maintaining its food security.

Given Thailand's reputation as one of the world's top producers of agricultural products, one would think that the country's bedrock must be its nutrient-rich soil. Not so, says Chalermpol Kirdmanee, principal researcher and head of Plant Physiology & Biochemistry Laboratory in the Agricultural Biotechnology Research Unit at the National Center for Genetic Engineering and Biotechnology (BIOTEC), the country's research arm on biological science. "Soil conditions in Thailand are like famished, malnourished kids who eat too much junk food", he said. It's a provocative way of saying that Thai soil has five times less the vital nutrients like nitrogen than the accepted global standard, which makes much of the land unsuitable for farming.

Across the nation, there is quite a range of

> "Soil conditions in Thailand are like famished, malnourished kids who eat too much junk food."
> -Chalermpol Kirdmanee, principal researcher and head of Plant Physiology & Biochemistry Laboratory in the Agricultural Biotechnology Research Unit at the National Center for Genetic Engineering and Biotechnology

Desertification
A type of land degradation in which a relatively dry region becomes increasingly arid, typically losing its bodies of water, vegetation and wildlife. It is caused by a variety of factors, such as climate change and man-made activities.

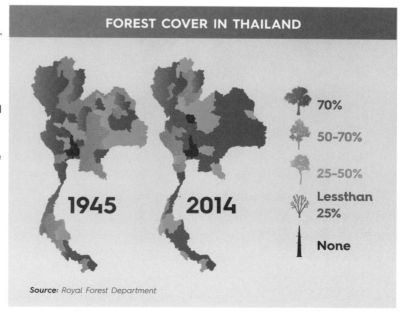

FOREST COVER IN THAILAND

1945 | 2014

- 70%
- 50-70%
- 25-50%
- Less than 25%
- None

Source: Royal Forest Department

different kinds of soil. In the north of Thailand and to the west bordering Myanmar there are highlands of sandy soil. The central plains consist of low-lying farmland characterized by alluvial soils fed by five rivers from the north, with mudflats at the mouth of the Gulf of Thailand. The gulf-facing parts of the Eastern Seaboard and southern provinces are filled with saline soil, or alkaline soil, unsuited to crop cultivation. And in the famously arid Northeast, prone to desertification, salt is also an issue as swaths of the region sit atop enormous salt deposits buried under the Korat Basin and the smaller Sakhon Nakhon Basin

Farmers in the northwest seed soy beans in burnt soil after a rice harvest.

(both were seabeds in prehistoric times).

Chalermpol, who has spent more than 20 years working with royally initiated projects on soil salinization in the Northeast, explains that many of the problems are man-made. Ignorance and greed top the list of culprits. Both are to blame for the monocropping that became a staple of local farms in the 1970s. This style of cultivation uses the same plot of land to plant the same type of crop over and over again until the nitrogen and other essential elements are leached from the soil. The intensive use of chemical fertilizers since the 1970s has also had a harmful effect on Thailand's soil, while modern farming equipment like tractors has compacted the earth, preventing moisture and organic matter from penetrating it. The damage has been devastating to say the least. According to the latest survey by the Land Development Department (LDD) under the Ministry of Agriculture and Cooperatives, more than 54 percent of the total land in the country is low grade, while the amount of **soil organic matter (SOM)**, a key indicator of the soil's health and fertility, is too low nationwide. Among these substandard varieties is acidic soil – meaning soil with a pH level lower than 5.5 – which accounts for 29 percent of soil in Thailand. This is mostly caused by the misuse of chemical fertilizers

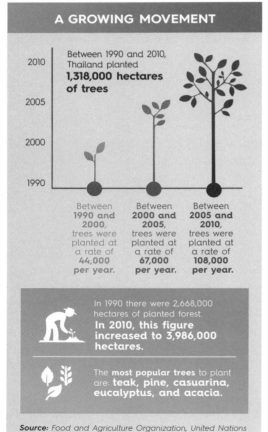

A GROWING MOVEMENT

Between 1990 and 2010, Thailand planted **1,318,000 hectares of trees**

Between 1990 and 2000, trees were planted at a rate of **44,000 per year.**

Between 2000 and 2005, trees were planted at a rate of **67,000 per year.**

Between 2005 and 2010, trees were planted at a rate of **108,000 per year.**

In 1990 there were 2,668,000 hectares of planted forest. **In 2010, this figure increased to 3,986,000 hectares.**

The **most popular trees** to plant are: **teak, pine, casuarina, eucalyptus, and acacia.**

Source: Food and Agriculture Organization, United Nations

"An intact forest cycles nutrients, regulates climate, stabilizes soil, treats waste, provides habitats, and offers opportunities for recreation."

-Janet Larsen, the Worldwatch Institute

and by the air pollution created by burning coal and bunker oil (a sludgy extract). The LDD also estimates that about one-third of Thailand suffers from soil erosion as a result of chronic land mismanagement.

Globally, Thailand keeps good company in facing these issues. According to figures released at the World Economic Forum of 2012, some 40 percent of the planet's soil is degraded. The main causes have been identified as high-yield agricultural practices requiring the intensive use of chemicals and over-plowing, which strips away topsoil, in addition to removing crop stubble by burning or animal grazing. At this rate of loss, the world's topsoil, which is vital for sustaining plant life, will be gone within 60 years.

Down the road, Thailand's primary challenges lie in employing sustainable means to combat desertification and restore degraded soil. Getting more carbon back into the soil through the elimination of bad farming practices like overgrazing, over-plowing and burning off crop stubble is one way to reinvigorate soil. Practices like monocropping and over-use of chemical fertilizers also need to be nipped in the bud. The most productive way to do this would be through outreach programs that educate farmers and offer them better, more sustainable alternatives. Proper land zoning could also help farmers find crops that better suit the soil conditions in their respective

Soil organic matter (SOM)

The organic matter component of soils, consisting of plant and animal residue, cell and cell tissues of decomposed soil organisms. SOM is a proven indicator of soil quality.

Solving Nature's Dilemmas from the Ground Up

While most monarchs are associated with opulently appointed palaces, those of King Bhumibol Adulyadej in Bangkok and Hua Hin have long served as labs and launching pads for environmental projects, as have the six Royal Development Study Centers dotted around the country. But soil conservation is not a mere hobby. The king spent so many hours working on issues of soil that his birthday, December 5, is now celebrated as World Soil Day by the United Nations.

One of the most groundbreaking initiatives was the Land Development Project in Phetchaburi province. Created in the 1960s, the project is renowned as the first to deal with land management through the introduction of proper irrigation systems into this drought-prone area as well as the application of organic fertilizers and soil treatments.

Establishing this agricultural cooperative in the village of Hoob Krapong was only the start. From there, the king and his team set up soil treatment projects across the country. In Chachoengsao province, for example, the arid land with salty soil became ground zero for a pilot program designed to show how micro-reservoirs can be used as a "water buffer" to slow down the spread of alluvial deposits brought about by rivers and streams in floodplains.

Over the years, more pilot projects took off. In Narathiwat province, the Pikun Thong Royal Development Study Center broke ground by using different techniques to "trick the soil" into becoming more fertile and less acidic. Across the Northeast, notorious for its salinized soil, dozens of projects have been implemented using the king's belief that solving nature's dilemmas requires natural solutions. That was the modus operandi of collaborative efforts between SCG and the Crown Property Bureau Foundation to launch more soil treatment projects that used saline-resistant plants and trees to siphon off saltwater from the ground.

"For the king, soil conservation is a philosophy in which he uses nature to control nature. He will look at the big picture and then use his understanding of science and local culture and traditional wisdom," explained Dr Chalermpol Kirdmanee of BIOTEC.

Alliances Between Villagers and Authorities Preserve Nature

In a community forest in the Northeast.

Over a million people live in and depend on Thailand's forests. Not only do forest-dependent villagers know wooded terrains the best, they also rely on forests for their livelihoods, making their protection a top priority.

While the Thai government recognizes the role villagers can play in forest preservation, critics say it has not gone far enough in empowering them to do so. Neither the National Park Act nor the National Forest Reserve Act of the 1960s, which set aside protected areas and suddenly turned those who had lived on the land for generations into trespassers, recognized the rights of locals to participate in decision-making processes regarding how to manage these areas. In a speech in 1973, King Bhumibol Adulyadej acknowledged the problem and stated: "They [the villagers] have human rights. It's a case of the government violating the people, not the people violating the law." It's no wonder that this state-managed approach did nothing to stop rampant deforestation, as the amount of forested area decreased from 53 percent in 1961 to 25 percent in 1998. When the government did finally realize the importance of involving local communities, the latter figure rose to around 28 to 30 percent.

Across Thailand, there are now 9,177 "community forests" registered with the Royal Forestry Department (RFD), where local residents have been empowered to manage these resources, sometimes collaborating with government agencies, civil society groups and even Buddhist monks. Experts believe that many of these projects have made a significant contribution to the maintenance of healthy forests, although the effect on local incomes and poverty reduction has not been as great as hoped for thus far.

Ronnakorn Triraganon of the Regional Community Forestry Training Center for Asia and the Pacific (RECOFTC), an NGO headquartered in Bangkok, points out that if local residents reap rewards for their conservation efforts it "provides the motivation for them to take care of the forest." As an example, he pointed out that today some communities earn up to two million baht a year from selling products such as bamboo, wild fruits, vegetables and honey.

Overall, positive examples of community forest management abound. One such project is the Joint Management of Protected Areas (JoMPA) based in one of the country's most biologically rich areas, the Western Forest Complex, where the Seub Nakhasathien Foundation has helped to foster cooperation between forest-dependent villagers and government officials. After winning the trust of both sides, the foundation equipped them with GPS systems which enabled villagers and authorities to survey the area and determine together which lands could be used for agricultural purposes and which lands would have to remain protected forest.

Meanwhile, the "Tree Bank," launched in 2006, has encouraged farmers to plant trees that they can use as collateral for low-interest loans. And from small beginnings in northern Phayao province in the late 1980s, the concept of "ordaining" trees in a Buddhist ceremony that entails wrapping them with sacred sashes as a way of protecting them from loggers, has spread across Thailand and to neighboring countries. The many projects focused on building check dams across the country have also become contributors to forest conservation. All in all, the battle to preserve Thailand's forests for future generations will not be easy, but as these projects show, it's far from a lost cause. •

regions. Scaling up the planting of vetiver grass in areas prone to erosion, an initiative that was first championed by King Bhumibol, would also go a long way toward mitigating the loss of fertile topsoil.

SEEING THE FOREST FOR MORE THAN JUST ITS TREES

We depend on forests for food, water, oxygen, medicines, raw materials and more. Some 80 percent of all plant and animal species live in the Earth's forests, which are also one of the best natural "technologies" we have for combating climate change. That's because forests

> "Deforestation accounts for some 10 percent of greenhouse gas emissions around the globe."

are **carbon sinks**. Using energy from the photons of the sun's light and combining it with carbon dioxide, forests create carbon and store it in their wood, leaves, roots and the soil surrounding them, ensuring that it doesn't escape into the atmosphere.

The good news is that Thailand has a wealth of forest cover: both evergreen and deciduous. In the north, the latter species cling to the highlands and are made up of such subspecies as cloud forests. In the northeast, dry dipterocarp and dry deciduous forests make up much of the tree line, whereas evergreen forests enrich the provinces of the south and the east. On the isolated fringes of the Eastern Seaboard and the southern flanks of the Andaman Sea are the vital yet much-decimated mangroves.

The bad news is that all of these are threatened by deforestation. Until the late-19th century this was not an issue in Thailand. That changed with the arrival of the British who were already carving up the teak forests of Burma as the fodder for beautiful hardwood furniture and the raw material for the ships that kept their empire afloat. These days, logging is well down the list of deforestation drivers in Thailand. Instead, commercial agriculture and subsistence farming vie for the top spots, followed by mining, major infrastructure projects like dams and roads, and urban encroachment. Drivers like these also help explain why deforestation accounts for some ten percent of greenhouse gas emissions around the globe.

In developing countries the root causes behind deforestation are not difficult to nail down: short-term profits trump the long-term benefits of human health and ecological welfare. It is not only large landowners who clear forests. In Thailand, unable to secure title deeds, for decades small-scale farmers have headed into the forests to clear plots for cultivation. These days, agribusinesses, for instance, encourage northern farmers to clear trees to grow corn and other monocrops. The

Carbon sinks

Rivers, forests, soil and the atmosphere are carbon sinks, areas that absorb and store carbon for long periods of time, thereby taking it out of the atmosphere.

Pollution from the burning of fields, like this sugarcane plot in the Northeast, is a growing issue.

Integrated forest management at the Huai Hong Khrai Royal Development Study Center project.

firms offer a good price for the crops, which are used as animal feed. The issue has caught the attention of the public and watchdogs, who decry the pollution that has resulted in the North from the burning of farms after the crops have been harvested. The public is now demanding that the companies supporting the farmers take more responsibility. Natural landscapes in the North in Nan and Chiang Mai provinces, for example, are increasingly being blighted.

As David W. Pearce summed up in his paper, "The Economic Value of Forest Ecosystems", "rapid population change and economic incentives...make forest conversion appear more profitable than forest conservation." This philosophy is endemic to most, if not all, of Southeast Asia. In Thailand's case it has had some disastrous consequences. According to the latest government statistics, Thailand's forest cover is roughly around 31 percent, though some experts insist it may be closer to 27 or 28 percent. The only way to get these statistics is through satellite images, which take about two to three years to compile. Whatever the figure, it still represents a steep decline from 1945 when around 60 percent of the forests were still standing. Though the rate of deforestation has slowed to around 0.2 percent per year on the low end, the implications are still dire, with floods, landslides and soil erosion taking their toll on communities and crops, while the decline of keystone species such as the tiger and hornbill threatens the country's biodiversity.

Thailand has actually been working to restore degraded forest areas since the 1970s, but initial efforts got off to a misguided start. Early government schemes focused on planting single cash crops, like eucalyptus, oil palms or rubber trees, however this practice failed to address the need to restore biodiversity to woodland areas. These plantations also have a detrimental effect on ecosystems and biodiversity (oil palm plantations, which have been expanding in Thailand at roughly nine percent per year, are of particular concern).

However, royal projects later initiated by King Bhumibol have been a catalyst for proper reforestation efforts. One of his strategies, known as "three forests, four benefits," recommends planting diverse forests instead of monoculture plantations. To celebrate His Majesty's Golden Jubilee in 1996, a plan was implemented to reforest more than 800,000 hectares of denuded land nationwide. Since then, many such efforts have favored King Bhumibol's approach over monoculture plantations, which are neither able to function as effective carbon sinks nor provide the full spectrum of benefits that a diverse forest can.

In particular, the Huai Hong Khrai Royal Development Study Center project, established in the Khun Mae Kuang National Forest Reserve in 1982, has been a paragon among integrated reforestation initiatives. Upstream, mountainous areas were reforested with a variety of trees and slopes were planted with vetiver grass to help the land retain moisture and prevent water runoff and soil erosion.

> *Thailand's Forest Cover* IS ONLY ABOUT 31 PERCENT

More trees were planted in the degraded forest, while downstream eight large reservoirs that hold approximately 3.3 million cubic meters of rainwater were constructed. In addition smaller reservoirs, check dams and small channels were built throughout the area to distribute moisture, slow down run-off, and keep fertile alluvial soil from being washed away.

Another bright spot has been the spread of **community forestry**, which allows villagers who live in and depend on these ecosystems to use their homegrown wisdom to manage them in more sustainable ways.

Reforestation has also risen as a priority under the current government, which hopes to increase forest cover to 40 percent over the next decade. To further this aim, the state has begun reclaiming land listed as forest reserves

> "The prospect of biodiversity in Thailand is very much in crisis. The present conservation efforts have not been able to hold back the alarming rate of biodiversity loss."
> –From a report by Thailand's Office of Natural Resources and Environmental Policy and Planning

from villagers, resort moguls and plantation owners. In the northeastern province of Sakon Nakhon, officials have taken back about 2,512 hectares of land that will be used to reforest parts of Phu Phan National Park and Phu Pha Lek National Park. While applauded for the intent to restore these forests, the plan has drawn the ire of some for its lack of consultation with local communities.

Overall, one major stumbling block is that Thailand's reforestation movement is largely fragmented, and lacks focus. And while Thailand has an impressive list of laws on the books to protect forests, enforcing such laws or prosecuting those who break them has proven challenging over the years. Additionally, the government is regularly lax about conducting transparent Environmental Impact Assessments (EIAs) for projects in woodland areas.

That said, moving forward, it is imperative that Thailand rigorously enforce its forestry laws and apply an integrated approach to conservation if it is to successfully halt deforestation and scale up efforts to restore degraded forests. To achieve this, three other questions should be taken into consideration:

How will Thailand mobilize adequate resources to finance and implement the kind of large-scale reforestation programs needed? Secondly, how will it enforce the long-term protection and sustainable management of forest areas? Thirdly, how will local communities be included and consulted to ensure that they too have a stake in forest conservation?

It's a tall order. But the answers to these questions will define Thailand's forest conservation record for years to come.

SAFEGUARDING A TREASURE TROVE OF BIODIVERSITY

Wherever you go in Thailand it is virtually impossible to ignore the country's incredible biodiversity. Indeed, scientists estimate that

Community forestry

Coined back in the 1970s and popular in places like North America, Brazil and India, community forestry describes how local villagers work with NGOs and the state to manage the forests near where they live. It's still a relatively recent trend in Southeast Asia.

PRICES FOR ILLEGAL WILDLIFE PRODUCTS

One kg of tiger bone
US$1,000 – 1,500

Tiger pelt
US$2,000

A live baby elephant
US$30,000

Worked ivory
US$5,000 per kg

Pangolin
US$320 per kg or around US$800 per whole animal

Slow loris
US$650 for a pet

Source: Freeland Foundation

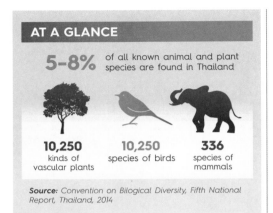

AT A GLANCE

5–8% of all known animal and plant species are found in Thailand

10,250 kinds of vascular plants

10,250 species of birds

336 species of mammals

Source: Convention on Bilogical Diversity, Fifth National Report, Thailand, 2014

between five to eight percent of all known animal and plant species on Earth are found here. This diversity is due, in part, to the country's location at a biogeographic crossroads between the Indochinese region in the north and the Sundaic region to the south. The problem is that this biodiversity is under constant threat from deforestation, urban encroachment, pollution, the poaching of exotic species and climate change.

In some ways, it could be argued that a loss of flora and fauna is inevitable given Thailand's rapid transformation from a largely agrarian nation to a Newly Industrialized Country. Starting in the 1960s, vast tracts of jungle were cleared to make way for commercial crops, such as rice, cassava and sugar cane. The resulting impact of deforestation on wildlife populations has been startling. Elephants, which numbered in the range of 100,000 individuals at the beginning of the 20th century, have fallen to around 3,000–3,500 in the wild today, while the number of tigers in the kingdom has been estimated at less than 250.

While poaching remains a serious issue, there has been some progress of late. In particular, there are signs that Thailand's wild tiger population is ever-so-gradually rebounding. According to a new study by the Wildlife Conservation Society (WCS), an intensive SMART Patrol system involving nearly 200 anti-poaching rangers has helped the Huai Kha Khaeng Wildlife Sanctuary acquire the status of the only site in Southeast Asia where researchers have confirmed tiger populations are growing. The joint initiative between WCS and Thailand's Department of National Parks, Wildlife and Plant Conservation (DNP) was set up in 2006 to curb the poaching of tigers and their prey, and to recover what is possibly the largest remaining "source" population of wild tigers in mainland Southeast Asia. As of 2016, the program has identified 90 tigers in Huai Kha Khaeng Wildlife Sanctuary alone.

However, the illegal wildlife trade continues to flourish and researchers warn that the threat posed by poaching isn't going away anytime soon. High prices for rare animals and plants drive a fast-growing black market for endangered species in Thailand and abroad. Sadly, the kingdom remains one of the major wildlife trafficking centers in Asia.

As threats to Thailand's endangered species and biodiversity continue to grow, one of the key long-term solutions is to encourage

Tipping point

The term refers to the critical juncture at which the collapse of biodiversity could irreversibly change life on Earth and threaten the survival of many species, including humans.

Anti-poaching rangers patrol in the Huai Kha Khaeng Wildlife Sanctuary.

Thailand's overall tiger population has declined significantly in recent decades but in Huai Kha Khaeng Wildlife sanctuary they are making a slow come-back.

local communities to play a greater role in wildlife protection. From promoting community awareness about the value of wild animals in the ecosystem to assisting former poachers to develop alternative sources of income, there are plenty of ways the government and NGOs can help. But the most important thing is for locals to gain an understanding of why they should protect wild animals, and how their own efforts can contribute to a sustainable future.

A shining example of this kind of community engagement occurred in Wang Mee district, just a short distance from Khao Yai National Park. Fifteen years ago, the area was a hotbed of poachers and illegal loggers. Today, thanks in part to the support and technical advice of the Freeland Foundation, the inhabitants have found alternative livelihoods such as growing organic mushrooms and mulberry trees. Although some have been lured back into the illegal trade by sky-high prices for endangered species, the combination of poverty alleviation programs and strict enforcement has led to a reduction in poaching levels of up to 75 percent in some areas.

Sustainable wildlife conservation requires strong collaboration between government agencies, NGOs, community-based organizations and the private sector. It also necessitates a clear plan of action, which luckily, Thailand has already laid out. The country's National Biodiversity Action Plan, covering the period from 2015–2021, calls for a reduction in the rate of habitat loss by 50 percent, the improved protection of threatened species, and a concerted campaign to better educate the public about the importance of preserving biodiversity throughout the nation. With the necessary tools now at its disposal, Thailand just needs to bridge the gap between good intentions and practical implementation to make these goals a reality.

The good news is that endangered species do have the potential to recover, so long as conditions are right. But Thailand needs to act fast, and decisively, to preserve its dynamic cradle of biodiversity before it reaches a **tipping point**.

Further Reading

- *World Wide Fund for Nature Thailand Annual Report, 2015*

- *Cost of Implementing National Biodiversity Strategies and Actions in Thailand: A Conceptual Approach*, UNDP, 2016

- *Parks for Life: Why We Love Thailand's National Parks* by Songtam Suksawang and Jeffrey A. McNeely, Department of National Parks, Wildlife, and Plant Conservation and UNDP, 2015

16 PEACE, JUSTICE AND STRONG INSTITUTIONS

Promote peaceful and inclusive society for sustainable development, provide access to justice for all and build effective, accountable and inclusive institutions at all levels.

PEACE, JUSTICE AND STRONG INSTITUTIONS

Sustainable Development Requires Dialogue and Peace

Calls to Action

- Find long-term democratic solutions to bridge the country's deep political divide; help opposing factions reach common ground
- Strengthen the integrity, accountability, effectiveness and transparency of institutions
- End violence, achieve reconciliation in the Deep South
- Guarantee equal access to justice for all; end impunity
- Prevent manipulation of laws and the justice system
- Stamp out endemic corruption and develop stronger, impartial monitoring mechanisms
- Protect human rights defenders; guarantee basic rights including freedom of speech and freedom of expression
- Ensure responsive, inclusive, participatory and representative decision-making at all levels
- End abuse, exploitation and trafficking of migrants

Thai parliament during a no-confidence vote in Bangkok in 2008. In the past, allegations of corruption have frequently led to instability, coups and new elections.

We inhabit a world that is increasingly divided. Across the globe, conflicts fueled by a variety of factors are perpetuating a level of instability that is antithetical to sustainable development. Such conflicts not only take a grave toll in terms of human life, but also lead to environmental degradation, consume valuable resources, weaken governing systems and prevent people from accessing healthcare, education, food, justice and enjoying other basic rights. Meanwhile, corruption and distortion of the rule of law pose major obstacles to progress and prosperity by undermining our institutions, stifling economic development and fueling political unrest.

Of all the SDGs, Goal 16 – which tasks policymakers with addressing these very issues in a sustainable way – arguably represents Thailand's greatest challenge. Certainly, the kingdom's progress on this front will greatly influence its ability to meet the other 16 SDGs as well. Here's why: In recent years, Thailand has been seeing an unprecedented polarization of its society. A breakdown of dialogue between opposing political factions has left democracy in the lurch and many of the kingdom's state institutions have weakened in the face of a variety of forces, ranging from a lack of policy continuity to political interference. Meanwhile, perennial issues such as the weak rule of law and corruption remain. As a whole, these detrimental forces threaten to derail Thailand's best efforts to chart a sustainable development pathway.

"Dealing with the symptoms will no longer be adequate to heal the gaping wound in our country. We need to deal with the bacteria and the viruses that lie at the roots of our national malaise", long-time politician Surin Pitsuwan wrote in a June 2016 *Bangkok Post* editorial. The statesman went on to warn: "It is high time that the Thai bureaucratic system be overhauled before a rebellion of the periphery."

Although it has been hundreds of years since Thailand was truly engulfed in war, it has not been immune to serious unrest. Since the abolishment of absolute monarchy in 1932, Thailand has been wracked by 12 coups d'état (and many other failed attempts). While the periods of military rule that followed have brought stretches of development, security and peace, they have also on occasion – in the 1970s, 1990s and 21st century – led to violent unrest.

As of 2016, Thailand suffers from two significant conflicts: one, between political factions typically known as the **red shirts** and **yellow shirts**, has resulted in frequent mass street demonstrations over the past ten years that have paralyzed large sections of Bangkok and triggered other tense protests across the country. This conflict dates back to 2005, when yellow-clad protesters opposed to then-Prime Minister Thaksin Shinawatra occupied large public spaces in Bangkok to call for his removal. After Thaksin was deposed in a 2006 coup, his supporters – hailing largely from the North and Northeast – donned red and took to the streets to protest his ousting. Over the next decade, this disruptive brand of political theater would play out time and again in the capital, sometimes with violent consequences. Most recently, months-long instability in 2013–2014 eventually resulted in the removal of then-Prime Minister Yingluck Shinawatra, a coup d'état and the establishment of a military

> **"Dealing with the symptoms will no longer be adequate to heal the gaping wound in our country. We need to deal with the bacteria and the viruses which lie at the roots of our national malaise."**
> –Politician Surin Pitsuwan

junta under the National Council for Peace and Order (NCPO), with General Prayuth Chan-ocha as prime minister. Elections have been forecast to take place in 2017.

The second conflict involves a localized Muslim insurgency in the three Deep South provinces of Yala, Pattani and Narathiwat – sometimes also spilling over into Songkhla province – where a large percentage of the region's population is of Malay descent and most residents speak Patani Malay (*Yawi* in Thai) as their first language. Since the flare up in 2004, more than 6,000 people have been killed, according to Deep South Watch, both

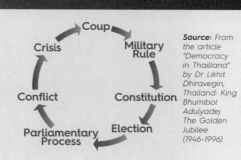

THAILAND'S POLITICAL CYCLE

Source: From the article "Democracy in Thailand" by Dr Likhit Dhiravegin, Thailand: King Bhumibol Adulyadej The Golden Jubilee (1946-1996)

Red shirts

Broadly speaking, most red shirts hail from the north and northeast. They may be backers of former prime minister Thaksin Shinawatra, be spurred by their perspectives on inequality or see the movement as a vehicle to achieve other objectives such as a redistribution of power or wealth.

Yellow shirts

They tend to be distrustful of Thaksin Shinawatra and his network and the aims of the red shirt leaders; they are generally comprised of supporters of the Democrat Party, royalists and the Bangkok middle class. They may also have benign objectives such as rooting out corruption and inequality, which seemingly overlap with the causes of the red shirts.

from insurgent attacks and as a result of military responses to the unrest. As historian and expert on the issue Duncan McCargo notes, the causes of the conflict are "complex and often opaque. Alternative explanations include questions of identity, historical injustice, economic inequality and discrimination, unequal power relations, and networks of criminality involving local politicians and members of the security forces."

In various ways, both of these conflicts have impacted the kingdom's image and hindered development. Thailand ranked 125 out of 163 countries in the Global Peace Index report of 2016. This index gauges global peace through 23 qualitative and quantitative indicators under three broad themes: the level of safety and security in society, the extent of domestic and international conflict and the degree of militarization. Only the Philippines and North Korea fared worse than Thailand in the Asia-Pacific region. This low ranking stemmed mostly from poor scores involving domestic issues such as "violent demonstrations", "terrorism impact" and a high rate of "violent crime". The report estimated the financial costs to Thailand at just over US$70 billion, which is $1,033 per Thai, or seven percent of GDP. Meanwhile, according to the SDG Index, Goal 16 was one of the few Goals in which Thailand scored lower than its region's average. Among the reasons were its very high prison population, high homicide rate and the high perception of corruption.

While Thailand's conflicts and crimes fill the newspapers and can give the appearance of a country in turmoil, the kingdom has remained incredibly resilient in many respects. In 2014 it ranked as the world's 14th-most-popular tourist destination, and although no stranger to slumps, it has maintained its status as the region's second-strongest economy. The corrosive effects of Thailand's conflicts are thus slightly harder to identify and the greatest loss is likely to be found in the unknown opportunity costs suffered over the long-term. In other words, a lack of policy and political continuity over the past decade in particular may have created an era of lost opportunities for the country during a time of increasing global competitiveness.

The ongoing nature of the unrest also means that Thai political institutions are forever channeling their energy into negotiating ways out of new crises, rather than into more long-term strategies of inclusive development. Moving forward, Thailand's challenge lies in transforming these conflicts into an opportunity to establish the kind of peaceful, constructive dialogue that has so far been absent between opposing sides, both in the Deep South and on the political battlefield.

RULE OF LAW AND CORRUPTION

Promoting the rule of law, safeguarding the independence of the judiciary and ensuring equal access to fair and equitable justice are key pillars of Goal 16. These precepts not only foster a healthy society, but also create the foundation needed for sustainable development to take root and flourish. Notably, according to a 2015 World Justice Project report, the

Thailand IS RANKED **125th** IN THE GLOBAL PEACE INDEX

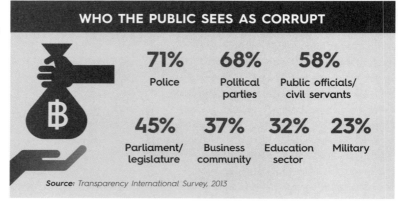

WHO THE PUBLIC SEES AS CORRUPT

71% Police
68% Political parties
58% Public officials/ civil servants

45% Parliament/ legislature
37% Business community
32% Education sector
23% Military

Source: Transparency International Survey, 2013

top four countries in terms of rule of law – Sweden, Norway, Finland and Denmark – are the top four countries in terms of sustainable development as ranked by the SDG Index. Thailand was 56th out of 201 countries ranked by the 2015 World Justice Project report and 61st in the SDG Index. Therefore, a clear correlation between rule of law and sustainable development appears to exist.

In this regard then, Thailand has much to improve if it is going to achieve Goal 16 by 2030. In a stark assessment during a 2016 speech to the Foreign Correspondents Club of Thailand, former prime minister Anand Panyarachun, said "…it has become patently clear that many of our institutions are inadequate when faced with the challenges of globalization. Against a backdrop of rapid global change, our economic, political, and social institutions have simply not kept up."

When General Prayuth Chan-ocha seized power, he declared from the outset his intent to reform state institutions with the ultimate aim of curbing corruption and nepotism. To date, numerous measures have been passed and actions, particularly on the issue of land encroachment, have been taken. However, as the process continues to drag out some stakeholders have criticized the way in which these reforms are being pushed through. For example, the prime minister has used Article 44 more than 50 times since taking power in 2014 to pass policies and to kick-start stalled reforms, according to iLaw, a Bangkok-based legal rights monitoring group. Article 44 allows Prayuth to issue any order in the name of national security, reforms, or unity regardless of the legislative, executive or judicial force of the order.

By and large, most Thais would agree that reform is necessary, with the eradication of corruption one of the desired outcomes. Thailand has long-struggled with a culture of

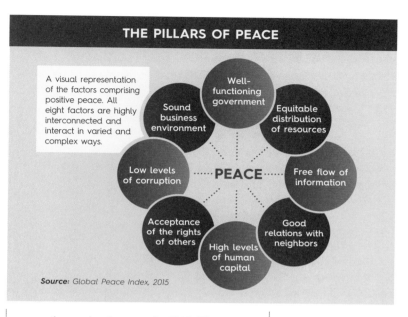

THE PILLARS OF PEACE

A visual representation of the factors comprising positive peace. All eight factors are highly interconnected and interact in varied and complex ways.

Source: Global Peace Index, 2015

corruption and patronage. In 2015, Thailand ranked 76th in the Corruption Perceptions Index among 168 countries, an annual report published by Transparency International, which is based on the perceptions of foreign businessmen. Overall, this kind of corruption can have a negative impact on the strength, credibility and impartiality of the nation's institutions, which in turn hinders sustainable development.

Corruption hits poor people the hardest, often having "devastating consequences", according to Transparency International. For example, a bribe demanded by a police officer may mean that a family can't afford school fees or even food to put on the dinner table. Corruption also means that the services people depend on – from drinking water to health clinics to roads – end up costing more, being of lower quality, or both. Corruption, bribery, theft and tax evasion actually cost developing countries about US$1.26 trillion per year, according to the United Nations Office on Drugs and Crime.

In general, Thais are fed up with corruption and people on both sides of the political divide view its eradication as one of the key issues driving their vote. That said, despite having several bodies to combat corruption – including a system of declaring assets and liabilities and an independent anti-corruption agency – corrupt practices are difficult to root out in Thai society, business and government. Even the anti-corruption institutions have suffered from corruption, and like the judiciary, have on occasion been manipulated to achieve political or financial ends. Indeed, corruption's impervious nature has so exasperated some that they've turned to the supernatural for help. In mid-2016, Thailand's public auditor office announced that in an effort to help safeguard public funds it would be distributing "magical scarfs" to people who are fighting against corruption.

DEMOCRACY AND INSTITUTIONS

"We have a tendency to focus on democracy in form rather than in substance," said Anand Panyarachun in his speech to the FCCT. "We follow procedures and go through the motions of elections. Yet we have paid little attention to developing the institutions that are critical to sustaining democracy. The challenges that we are presently facing have their roots in the fact that we have never had a true democratic transition – a genuine change in our political system."

Indeed, Thailand's already tenuous democratic institutions have grown even less stable since the mid-2000s. This includes a weakening and marginalization of political parties, the judiciary, the Senate, the Election Commission of Thailand, the National Anti-Corruption Commission and the Ombudsman. The party system is fragmented and highly polarized. Political parties in Thailand are created from the top down, and are largely driven by parliamentary, military or business elites. In general, these parties lack longevity, ideology, and tend to be clan-controlled and fueled by clientelism.

> "We have a tendency to focus on democracy in form rather than in substance. We follow procedures and go through the motions of elections. Yet we have paid little attention to developing the institutions that are critical to sustaining democracy."
>
> -Former prime minister Anand Panyarachun

As Pasuk Phongpaichit and Chris Baker noted in their 2016 book *Unequal Thailand: Aspects of Income, Wealth and Power*, "…the weak development of Thai political parties is a result of the general immaturity of Thailand's political system." Meanwhile, they continued, the "parliament is like an infant who, as soon as she grows up a little and shows some independence, is slapped down and put back in the cradle, and hence never gets a chance to become more mature."

In terms of governance, Thai technocrats have long called for decentralization, advocating a comprehensive redistribution of powers focused on empowering stakeholders at provincial and local governance levels. But under the current bureaucracy, every official – junior or senior – is appointed by Bangkok, in all areas of government administration. "In the

Thai police stand in front of the King Rama V Equestrian monument during a protest in Bangkok.

Keys to Bridging Divisions in Thailand

SINCERE DIALOGUE Whether between red shirts and yellow shirts, security forces and militants, or community stakeholders and local leaders, sincere dialogue absent of self-interest and personal agendas and bias is the only way inclusive and sustainable solutions will be found.

A FREE, UNBIASED PRESS Thailand ranked number 136 out of 180 countries in a Reporters Without Borders survey on press freedom for 2016. A free and objective press could provide a platform for airing the views of opposing sides on national issues, which in turn would go a long way toward healing rifts.

ON THE FRINGES Malay-Muslims in the Deep South have long complained that they feel detached from mainstream Thai society and culture. Recognizing that the issue is a cultural one as well as a security one would undermine local support for separatist groups and help pave the way for peace in southern Thailand.

RIGHTING HUMAN WRONGS Security forces and insurgents in the Deep South must stop committing human rights abuses, fanning the flames of the conflict.

REDISTRIBUTING WEALTH A disproportionate amount of government revenue is spent in Bangkok. Decentralization would help develop poorer regions of the country and address lingering resentment toward the capital's elite that has galvanized the red shirts.

RULE OF LAW Implementing a culture that respects the rule of law would deter protesters from resorting to illegal means to achieve their goals and demonstrate that negotiation, rather than brinkmanship and violence, is the way forward.

INVESTIGATE DISAPPEARANCES AND KILLINGS Setting up a special commission to investigate the killings and disappearances of human rights defenders would demonstrate a commitment to the rule of law.

---

past, we have established local institutions but always retained centralized control over them. Going forward, we must reform these local bodies so that they become answerable to the needs and demands of the local populace rather than to the central government," Anand said.

The Bank of Thailand (BOT) has, over the past decade or so, been one of Thailand's most dependable state institutions. After failing to safeguard the country's economic house during the 1997 financial crisis, the BOT recognized that it needed to put into place procedures not only for its own conduct, but also for regulatory and supervisory reforms for the country's financial system as a whole. Over the years, these changes have resulted in stable growth, lower inflation, lower-than-average interest rates, a relatively stable currency and sound foreign exchange reserves even at times of severe cyclical stress and volatility in the global economy.

Overall, Thailand's banks and other financial institutions remain strong. In May 2016, Moody's Investors Service forecast a stable outlook for the banking system over the following 12–18 month period, reflecting strong loss-absorption buffers within the country's banks.

For better or worse, the military is one of Thailand's strongest institutions. For decades, as the country experienced numerous political changeovers, the Army has served a pivotal role as peacemaker, buffer and facilitator of change. However, Thais cannot continue to rely on an armed group to play referee to democracy. Like the rest of the country, the military is riven by factions. And while effective at clearing the streets of protesters, military intervention does not equate to long-term stability.

While the current government has vowed to overhaul the nation's institutions and is drafting a new constitution, monitors like the Asia Foundation have criticized the process for its lack of "public input or consultation". It should be noted that Goal 16 strongly emphasizes the premise of promoting "inclusive, participatory and representative decision-making."

THE CHALLENGE OF PROMOTING A JUST AND INCLUSIVE SOCIETY

In recent years, in an attempt to stop turmoil in Thai society, press freedoms and free speech have been curbed significantly, while journal-

Countering Corruption in the Kingdom

Thailand's private sector has long been aware of the malaise of corruption and the toll it takes on their economic fortunes. On many occasions, corruption has derailed major investments and this has negatively impacted the country's competitiveness, as shown in recent surveys which forecast that future opportunities will be lost to other ASEAN nations less prone to graft.

Fortunately, lessons learned in other countries – particularly in Hong Kong and South Korea – of proven successes in tackling corruption to boost economic growth have been widely shared among large and small enterprises in Thailand, leading to the birth of Thailand's Private Sector Collective Action Coalition against Corruption (CAC) and the Anti-Corruption Organization of Thailand (ACT). Since the CAC's inception in 2010, almost 180 companies have been certified for their implementation of effective anti-corruption policies; a few of the stauncher enterprises have even declared "zero tolerance" for such misconduct. By mid-2016, 724 private companies had declared to the CAC their intentions to run "clean" businesses.

To aid this anti-graft drive, the Thai Institute of Directors (IOD), a founding member of the CAC, offers anti-corruption courses to disseminate best practices to a wide range of enterprises. Thus far, some 20 training sessions have taken place, involving more than 400 executives and chief compliance officers.

In this gradual way, good corporate governance is becoming a priority for companies that see the rewards it brings, like being listed on the Dow Jones Sustainability Indices. In 2016, 15 Thai companies made the list, the highest number out of any ASEAN state. For years, the indices have produced key indicators for investors to show which companies are excelling both ethically and environmentally and, in turn, which are best adapted to thrive in an investment climate that increasingly considers a firm's environmental, social and governance (ESG) standards. In 2015, 23 Thai-listed companies received the ASEAN Corporate Governance (CG) award and the kingdom's CG score of 75 percent was the highest within ASEAN.

In terms of financial institutions, the Stock Exchange of Thailand's (SET) partnership with the UN Sustainable Stock Exchanges Initiative has proven its seriousness about taking governance standards to the next level. In fact, the SET became the first stock market in ASEAN to join 12 others from around the world that are equally committed to promoting long-term sustainability by enhancing corporate transparency and ESG integration.

But to win this war the private sector cannot go it alone. Collaborating with other agencies like the National Anti-Corruption Commission (NACC), the Anti-Money Laundering Office (AMLO), and the ACT is crucial to investigating corruption and keeping it to manageable levels, if not eliminating it entirely.

ists and activists have faced state sponsored intimidation. Thai media can also display overt bias at times.

While Thailand has an active civil society, with more than 18,000 NGOs currently registered in the kingdom, hostility toward these groups on the part of the military, politicians, private businesses and other actors have hindered their capacity to operate freely on some levels. Under the current government of 2016, critics face the prospect of being sent to "re-education camps" if they step out of line on more than one occasion. According to iLaw, since the 2014 coup hundreds of people have been summoned by the junta for attitude adjustment.

Rights monitors note that Thailand's courts and laws have frequently been manipulated by powerful individuals, companies and political actors as a means to further their aims. Corruption-related prosecution has regularly been wielded as a political weapon by various administrations since the mid-2000s, and laws like *lèse majesté*, criminal defamation, sedition and the Computer Crimes Act are increasingly employed to silence dissenting voices, settle personal vendettas, eliminate political opponents or simply to make an example out of someone. Members of the news media, civil society activists, politicians and government critics are frequently targeted by such lawsuits. While this phenomenon is not endemic to Thailand alone, it clearly contradicts the aim of building an inclusive society that welcomes debate among parties with differing views.

Migrants also do not benefit from equal justice in Thailand. They are at risk of being scapegoated for crimes, targeted for bribes and exploited by employers. They have limited access to legal protection and are reluctant to report crimes due to fears of losing their jobs or having negative interactions with police. According to an Asia Foundation survey, even

A new generation of tech-savvy activists are taking their causes to cyberspace.

Thais see the police force as the country's least trustworthy institution, with 39 percent of those surveyed ranking the integrity of the police as low or very low.

In particular, Thailand's *lèse-majesté* law, which makes it a crime to insult the royal family, has seen a significant spike in use during the latest period of political upheaval. As outlined by Section 112 of the Criminal Code, "Whoever defames, insults, or threatens the king, the queen, the heir-apparent, or the regent, shall be punished with imprisonment of three to fifteen years." While it is common for countries with constitutional monarchies to have such laws, in Thailand their application is sometimes perverted by individuals and political actors with ulterior motives.

Since the May 2014 coup d'état, the new government has aggressively pursued *lèse-majesté* cases, charging at least 68 people under Section 112. Around 50 people have also been charged with sedition (Section 116 of the Criminal Code) since the coup. Under martial law, which was established across the country following the coup, all such cases including those involving civilians are tried in military courts, a practice that watchdogs say violates international fair trial rights. "While

> "The weak development of Thai political parties is a result of the general immaturity of Thailand's political system."
>
> -Pasuk Phongpaichit and Chris Baker, authors of *Unequal Thailand: Aspects of Income, Wealth and Power*

A Thai soldier helps children off a school bus in the deep south. Militants in the area have targeted schools, which they see as representative of the state.

insisting they aren't dictators, the Thai generals have used the military courts as a central feature of their crackdown against peaceful criticism and political dissent," said Brad Adams, Asia director of Human Rights Watch.

Even outside Thailand's political sphere, the weak implementation of rule of law has had a detrimental effect on the distribution of justice. Thai society is still very much governed by a hierarchy of power and position. From police to tycoons to village headmen, those seen as having elevated status are treated with deference. To challenge that hierarchy in a brazen manner can provoke serious legal action initiated by the more well-connected individual, or even violence.

In rural areas, those who take a stand against powerful individuals or companies are particularly at risk. Thailand is ranked the eighth most dangerous country in which to defend land and environmental rights, according to a 2014 report by Global Witness. Some 60 activists who led opposition to coal plants, toxic waste dumping, land grabbing or illegal logging have been killed or have disappeared under dubious circumstances in the past two decades, according to a May 2016 report in *The New York Times*.

Because many of these incidents occur in rural areas and are linked to well-connected individuals or high-profile business interests, these killings are unlikely to receive much media coverage and investigations tend to produce few meaningful convictions. Freedom House's 2015 country report on Thailand notes that "even in cases where perpetrators are prosecuted, there is a perception of impunity for the ultimate sponsors of the violence."

Some private companies that find themselves facing strong opposition from local communities will also try to bend the judicial system to their will. A mining company is seeking 50 million baht (US $1.4 million) in a civil defamation suit filed against a group of villagers in Loei province for taking part in a public event that called for the closure of the company's open-pit copper-gold mine and urged rehabilitation of the local environment.

The way forward won't be easy. Institutions serving the public need to cultivate ethical mindsets throughout the entire structure of their organizations and demonstrate a commitment to operating with integrity, and without bias. There needs to be more public participation and multi-stakeholder engagement. Checks against the abuse and manipulation of laws should be put into place, as should measures to guard against corruption and impunity. On an administrative level, the introduction of specialized courts would help to deliver efficient and timely justice on issues such as the environment and trafficking, and help cut down on the number of appeals handled by the Supreme Court. In addition, if a central agency such as the Justice Ministry was designated to oversee the law and legal systems it would go a long way toward promoting fair and equitable justice.

According to Anand Panyarachun, "Democratic governance ultimately is a state of mind rather than some tangible rule or procedure. Over and above the implementation of critical reforms, moving forward towards a prosperous new normal requires that we fundamentally change our way of thinking, attitudes and mindsets to embrace openness, a diversity of views, as well as values that support societal change. Democratic governance opens up channels through which the diversity in our society can come together to foster political, economic and social development. It thus represents the most direct route to true sustainability."

Thailand HAS MORE THAN *18,000* REGISTERED NGOs

Further Reading

• *Truth on Trial in Thailand: Defamation, Treason, and Lèse-Majesté* by David Streckfuss, 2011

• *Good Coup Gone Bad: Thailand's Political Developments since Thaksin's Downfall*, edited by Pavin Chachavalpongpun, Institute of Southeast Asian Studies, 2014

• *Anti-Corruption Strategies: Understanding What Works, What Doesn't And Why?*, UNDP, 2014

17 PARTNERSHIPS FOR THE GOALS

Strengthen the means of implementation and revitalize the global partnership for sustainable development.

PARTNERSHIPS FOR THE GOALS

The Strategies, Know-How and Financial Needs for the 2030 Agenda

Calls to Action

- Improve continuity of cooperation policies between Thailand and its international partners
- Encourage local ownership of development projects among beneficiary partners
- Enhance South-South, North-South and triangular cooperation on disseminating knowledge related to science, technology and innovation; further enhance capacity-building support to developing countries
- Bolster the development, transfer, dissemination and diffusion of environmentally sound technologies
- Increase imports from Least Developed Countries (LDCs) and adopt preferential rules of origin applicable to imports from LDCs
- Increase Official Development Assistance and Foreign Direct Investment in LDCs
- Promote effective public, public-private and civil society partnerships

Thailand is working hard to forge partnerships both at home and abroad.

The success, or failure, of the 2030 Agenda for Sustainable Development hinges upon our ability to cooperate effectively and forge sincere partnerships to address the daunting global challenges we face today. Issues like migration, climate change, food security and financial stability do not know national boundaries. Humanitarian crises brought about by conflicts and natural disasters continue to demand attention, while many Least Developed Countries (LDCs) desperately need Official Development Assistance (ODA) to close economic gaps, and spur trade and internal growth. Promoting investment, sharing technology and know-how, and helping these developing countries to increase exports and better manage their debt is also essential to fostering sustainable growth and development.

Broadly speaking, Goal 17 may appear to be the most abstract. But the multilateral agreements and regulatory frameworks, international conferences and cross-border economic initiatives which forge partnerships in terms of strategy, trade, policy coherence and commitment, as well as technical expertise and capacity building, are absolutely essential for the 2030 Agenda for Sustainable Development to see meaningful progress, let alone be achieved.

Critics wonder how the world could possibly fund such an ambitious development program. All countries must do their part, however small that may be. As a successful middle-income country with decades of experience in advancing human and economic development, Thailand is well positioned to be an important contributor to the kind of "global partnership" called for by Goal 17. Whereas Thailand was once primarily an aid recipient, it has now evolved to become a significant donor, trade partner, technical advisor and provider of Foreign Direct Investment (FDI).

The kingdom's global development approach is guided by SEP-inspired principles that emphasize the importance of human development, capacity building, fostering self-reliance and sharing lessons learned. With a strong commitment to assist its neighbors to achieve sustainability and prosperity, Thailand has been providing technical assistance and capacity support to less well-off countries in the region for more than five decades through the work of the Thailand International Cooperation Agency (TICA) and its many partners.

Reaching far beyond Southeast Asia, Thailand is also pursuing a "Look West" policy in an effort to promote constructive **South-South cooperation** through the sharing of experiences and best practices. In doing so, the kingdom plays an important role in bridging gaps among developing countries in the global south, and between developed and developing countries (through North-South cooperation and triangular cooperation). Disseminating its own expertise through various partnership initiatives, Thailand has collaborated with a

> "As a successful middle-income country with decades of experience in advancing human and economic development, Thailand is well positioned to be an important contributor to the kind of 'global partnership' called for by Goal 17."

South-South cooperation
When two or more developing countries from the Global South work together to achieve development through the sharing of resources, technology and expertise.

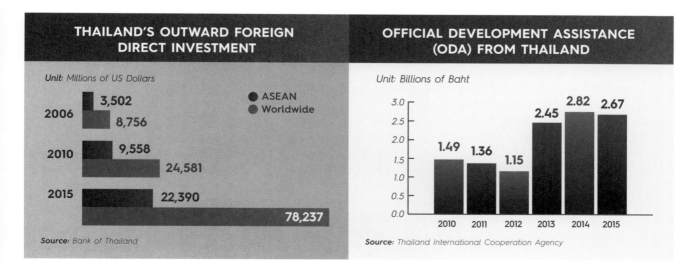

THAILAND'S OUTWARD FOREIGN DIRECT INVESTMENT

Unit: Millions of US Dollars

● ASEAN
● Worldwide

2006: 3,502 / 8,756
2010: 9,558 / 24,581
2015: 22,390 / 78,237

Source: Bank of Thailand

OFFICIAL DEVELOPMENT ASSISTANCE (ODA) FROM THAILAND

Unit: Billions of Baht

2010: 1.49
2011: 1.36
2012: 1.15
2013: 2.45
2014: 2.82
2015: 2.67

Source: Thailand International Cooperation Agency

range of developing countries on tackling issues such as poverty reduction, integrated farming, sustainable fisheries, conservation, HIV/AIDS, the narcotics trade, public health and human trafficking.

For example, Thailand has collaborated with Japan in Asia and Africa on the Third Country Training Programme, which provides courses in areas such as public health, reforestation, electricity system development, anti-human trafficking measures and survivor rehabilitation, environment and water resource management, food processing, malaria prevention and HIV/AIDS prevention. In Laos, joint Thai-German projects have helped to upgrade the paper mulberry supply chain in order to enhance rural cross-border economies, and worked to improve water and river quality in rural areas such as the Nam Xong Sub-River Basin. Meanwhile, Thai-French cooperation in the Mekong sub-region and Africa focuses on education/vocational training, agriculture and fisheries development, and public health (focusing on HIV/AIDS, malaria, tuberculosis and other contagious diseases). It is common for Thailand to partner with other Southeast Asian countries (i.e., Brunei, Indonesia, Malaysia, the Philippines and Singapore) to provide expertise to assist third countries on development initiatives.

Since 2013, the Thai-Africa Initiative has sought to enhance development and economic cooperation between Thailand and African countries by exchanging experiences, formulating policies directed at mutually beneficial trade and investment, and encouraging public, private and civil society partnerships. Additionally, Thailand operates the Thai Volunteer Programme, under which Thai experts work alongside their African counterparts to share

> THAI OFFICIAL DEVELOPMENT ASSISTANCE WAS *US$78 MILLION* IN 2015

Thailand has become more active in Africa in recent decades, helping Lesotho and other countries establish sustainable projects.

PARTNERSHIPS FOR THE GOALS

THAILAND COOPERATION AROUND THE GLOBE

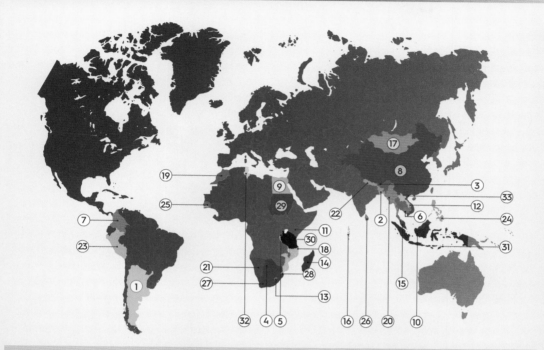

technical experience on projects in Africa.

Thailand also provides significant amounts of FDI and Official Development Assistance (ODA), most of which is earmarked for regional Least Developed Countries. In 2014, Thailand's ODA was estimated at around US$87 million, while in 2015 the figure was estimated to be around US$78 million, according to TICA. Thai ODA is based on a combination of Thailand's own development expertise and adaptations of suitable international techniques. As such, much of Thailand's ODA goes toward infrastructure development such as the construction of roads, bridges, dams, hospitals and power stations. The remainder is disseminated in the form of technical assistance and training in areas such as education, public health, agriculture, transportation, economics, banking, finance, and science and technology, as well as contributions to the UN System and the Asian Development Fund of the Asian Development Bank.

In general, Thailand places a strong emphasis on multilateral cooperation. Through the United Nations Partnership Framework (UNPAF) for Thailand, the Thai government and the United Nations Country team in Thailand (UNCT) are working toward promoting a just society, strengthening economic and security cooperation in the region, and sustainably managing natural resources and ecosystems.

In the area of International Cooperation, Thailand is expanding its capacity as a key international development partner, with TICA playing a major role in collaboration with the Asian Development Bank, the United Nations Development Programme (UNDP), the United Nations Population Fund (UNFPA), UNICEF and a number of international NGOs. Some of these include the Asia Foundation, Kenan Institute Asia, Konrad Adenauer Stiftung, Norwegian Church Aid, Stockholm Environment Institute and the Rockefeller Foundation.

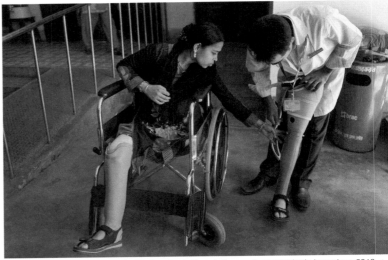

TICA has been helping Bangladesh's Prosthesis Foundation produce prosthetic legs since 2013.

In cooperation with the Prostheses Foundation, TICA has assisted Bangladesh in producing prosthetic legs since 2013, and in 2014 TICA conducted training courses on Lower Extremity Prostheses and Prostheses Evaluation for Bangladeshi healthcare professionals. Thailand also cooperated with UNDP to share learnings with Bangladesh about the One Tambon One Product initiative and how such programs can be used to foster community development.

Recognizing the importance that trade represents in helping LDCs to stand on their own economically, Thailand imports a great deal from its LDC neighbors, including agricultural products, labor-intensive manufactured goods, primary products, electricity and natural gas. In 2015, Thailand's imports from Myanmar totaled some $3.6 billion, while imports from Laos that year were around US$1.46 billion and imports from Cambodia were about $638 million, according to the World Integrated Trade Solution.

Thailand's status as a major manufacturing hub also makes it a popular destination for

Thailand's FDI in ASEAN COUNTRIES WAS US$22.4 BILLION IN 2015

migrants from neighboring countries who hope to increase their earning potential. By opening up its borders, Thailand not only contributes to its own GDP, but also toward empowering migrants from neighboring LDCs. Of the roughly 1.56 million migrant workers registered with the kingdom's Department of Labor, some 1.39 million are from Myanmar, Cambodia and Laos. Efforts to provide better oversight in the main industries that employ migrants is also helping to reduce exploitation and abuse.

Meanwhile, Thailand's private sector is increasingly investing abroad, primarily in agriculture, fisheries, food processing, construction, manufacturing, distribution centers, hotels and tourism. As of Q1 2016, Thailand had invested a total of US$58 billion in developing countries, 43.7 billion of which was in LDCs. In 2015, Thai companies invested some US$3.37 billion in Myanmar, $2.27 billion in Laos and $677 million in Cambodia, according to the Bank of Thailand. That same year, the kingdom's total FDI in ASEAN countries totaled almost $22.4 billion.

SHARING SEP EXPERTISE

Thailand sees the Sufficiency Economy Philosophy as a comprehensive solution to global sustainable development challenges, and as a result, SEP lays the foundation for most of Thailand's international development initiatives. In 2016, Thailand held the G77 Chairmanship, which it used in part to present the virtues of the Sufficiency Economy Philosophy in advancing the SDGs. A number of countries had by 2016 already incorporated Sufficiency Economy concepts into their respective development initiatives, including Cambodia, Indonesia, Laos, Lesotho, Myanmar, Timor-Leste and Tonga. In most cases, these SEP initiatives promote sustainable livelihood development and the use of integrated agriculture based on SEP and "New Theory" farming.

Through TICA, which is the primary focal agency in implementing international development cooperation between Thailand and

> **"Thailand sees the Sufficiency Economy Philosophy as a comprehensive solution to global sustainable development challenges, and as a result, SEP lays the foundation for most of Thailand's international development initiatives."**

developing countries, Thailand implemented a joint project with Lesotho during 2006–2010 to promote sustainable agriculture as an alternative and effective approach to rural development in the tiny African nation. The project focused primarily on setting up a demonstration center for sustainable agriculture based on SEP practices that have been used success-

Much of Thailand's SEP-related technical cooperation focuses on sustainable agriculture.

fully by Thai farmers. During the project, local farmers received on-the-job training from Thai experts in Lesotho and TICA organized a study visit program to Thailand for agricultural extension officers and farmers so they could learn alongside Thai farmers who have improved their quality of life by applying SEP.

In Timor-Leste, bilateral cooperation with Thailand since 2003 has led to the establishment of an SEP Model Village and Technology Transfer Center, which is geared toward enhancing the capability of government officers through agricultural education and technology training in crop production as well as promoting sustainable agricultural techniques that are in line with Timor-Leste's local needs and strengths. TICA also helped to develop a model village to demonstrate effective crop production based on SEP principles, and constructed a learning center accessible to farmers and the general public.

More recently, in February 2016, technical cooperation between Thailand and Tonga led to the launch of the Agricultural Development Cooperation Project, which aims at cultivating sustainable agriculture in Tonga based on SEP. The three-year pilot project will include growing fruits and will be extended to include initiatives such as fish farming.

TICA organizes Annual International Training Courses (AITC) related to SEP for individuals from more than 40 countries in Asia, the Pacific, the Middle East, Africa and Latin America, covering topics such as modern technology in sustainable agriculture, and the application of SEP in organic farming. In 2014, a total of 228 AITC scholarships in 14 courses were awarded through Royal Thai Embassies and Consulates around the world. Thailand also provides full scholarships for individuals from developing countries to study in SEP-related masters programs at Thai universities under the Thai International Postgraduate Program. In 2014, some 55 scholarships were awarded to officials from around the world to study in 22 different courses.

Additionally, Thailand regularly hosts the annual Buakaew Roundtable International Study Visit, during which participants (mostly from foreign affairs or international development cooperation backgrounds in developing countries) can see firsthand how SEP can be applied to address sustainable development challenges relevant to their own countries.

Finally, the kingdom also occasionally organizes one-off, country specific programs. For example, TICA organized a study visit for Colombian officials to learn about Thailand's experiences with poverty alleviation and rural development aimed at reducing development disparities between urban and rural areas.

Certainly, within Thailand there is much work to be done to achieve the type of multi-stakeholder cooperation that SEP and sustainable development require to succeed. Fortunately, the trends are positive. Increasingly, a consensus is growing and a mindset is being cultivated that respects the concept of balanced growth, and hopefully Thailand's commitment to sustainable development, buoyed by our framework of SEP, will continue to allow the kingdom to sincerely contribute to the 2030 Sustainable Development Agenda.

PARTNERSHIPS ON THAI SOIL

Within Thailand there are many organizations working toward helping the country achieve sustainable development, often forging partnerships that improve their collective ability to share knowledge and best practices. As is the case in Thailand's development work overseas, many of these initiatives are implemented bearing in mind the principles of SEP.

Countries pursuing SEP-inspired development initiatives with Thailand's assistance: Cambodia, Indonesia, Laos, Lesotho, Myanmar, Timor-Leste, Tonga

Thailand Sustainable Development Foundation (TSDF), as an example, dedicates the majority of its efforts to generating, collecting and disseminating knowledge that fosters sustainable development. As such, TSDF supports the projects and activities of public and private organizations, which share its concern for enhancing balance and sustainability in Thailand's economy, society, environment and culture — mirroring the four dimensions highlighted by SEP.

The Chaipattana Foundation, founded in 1988, assists rural people through community engagement with a focus on improving livelihoods. One such initiative, known as the Ampawa-Chaipattananurak Conservation Project, helped to revive and rejuvenate the Bangchang community along the Ampawa canal, a small tributary of the Mae Klong River in Samut Songkhram province. With the foundation acting as a planner and facilitator, the initiative focused on providing residents with the skills needed to improve livelihoods, build up the local economy and preserve the local culture as a way to achieve self-reliance and sustainability.

Meanwhile, the Thailand Development Research Institute (TDRI) focuses more on helping to enact policy change. TDRI provides technical analysis to various public agencies in an effort to help formulate policies that support long-term economic and social development in Thailand. In 2015–2016 alone, TDRI has helped to spread awareness on topics as diverse as support for disadvantaged mothers, good governance in public utility administration, labor law reform and ways to address shortfalls in the nation's education system.

Promoting sustainability and ethical consumerism has long been one of the main tenets of the Foundation for Consumers, Thailand's only national consumer advocacy

Youth volunteers help some of Oxfam's local partner organizations construct a daycare center for an ethnic Hmong community in Chaing Rai province.

agency that operates based on a participatory process. Over the long term, the foundation's watchdog function has helped keep companies in check, encouraging more ethical practices and upgrades in production standards by raising awareness among consumers and educating them to have higher expectations.

Numerous organizations in Thailand are also working on conflict resolution. In the volatile Deep South alone there are multiple organizations working on peace building such as Deep South Watch, the Muslim Attorney Center, Patani Forum, the Network to Uphold Justice for Peace, the Cross Cultural Foundation and the Duay Jai Group.

In addition to the many Thai organizations working on sustainability issues, there are numerous international agencies operating on Thai soil, collaborating closely with the Thai government and other local partners. The United Nations' various agencies such as UNDP, UNICEF and the International Labour Organisation have a strong presence in the kingdom as do NGOs such as Oxfam, the International Organization for Migration and the Asia Foundation.

Further Reading

• *Positioning the ASEAN Community in an Emerging Asia: Thai Perspectives,* compiled by the Department of ASEAN Affairs, Ministry of Foreign Affairs of the Kingdom of Thailand, 2016

• *Greater Mekong Sub-region Statistics on Growth, Infrastructure, and Trade,* by Asian Development Bank, 2016

A Call to Action: Thailand and the Sustainable Development Goals was based, in part, on our previous publication *Thailand's Sustainable Development Sourcebook* (2015). Some material from that volume was reproduced in this new work. *Thailand's Sustainable Development Sourcebook* featured contributions from editors, writers, readers and designers listed here, and was made financially possible thanks to the companies named on the opposite page.

Editorial Team

Alex Mavro (contributing editor) believes in business as a calling to serve the needs of others so that in the end, everyone profits. He is a highly regarded writer and speaker on the subject of purposeful stakeholder responsiveness and Chief of Operations, Centre for Sustainability Management, Sasin Graduate Institute of Business Administration of Chulalongkorn University.

Amornrat Mahitthirook (writer) is a 20-year veteran of the Bangkok Post who covers the transportation sector.

Amy Wu (writer) has over 20 years of professional journalism experience, including at TIME, San Francisco Chronicle, and The Deal.

Anchalee Kongrut (writer) has been a reporter at the Bangkok Post since 1997. She focuses on environmental issues. She has Master of Arts on mass media studies from The New School for Social Research in New York City.

Anjira Assavanonda (writer) spent 16 years reporting local and international stories with special focus on health and social issues for the Bangkok Post.

Apiradee Treerutkuarkul (editor and writer) spent the last 15 years as a print journalist and communications consultant for non-governmental organizations based in Bangkok prior to helping lead this project.

Aporntrath Phoonphongphiphat (writer) is a journalist with more than 20 years experience covering agricultural and industrial commodities.

Arianna Flores (writer) is a political scientist, who earned a Masters degree in Environmental Management and Technology at Mahidol University. Her recent research focuses on renewable energy, governance and the environment, and sustainable cities.

Ben Davies (writer) is a Bangkok-based journalist and photographer. He regularly contributes articles on the international wildlife trade and is author of Black Market - Inside the Endangered Species Trade in Asia.

Benjapa Sodsathit (art director) received an MFA from Minneapolis College of Art & Design in Visual Studies and a BFA from Silpakorn University. She is co-founder of Palotai Design Co., Ltd., which serves clients locally and internationally.

Chanthipapha Sopanaphimon (designer) earned her BFA from Silpakorn University.

Deunden Nikomborirak (writer) is a research director at the Thailand Development Research Institute, a private economic policy think-tank in Bangkok. Her area of expertise include competition policy, sectoral regulations, governance, anti-corruption strategy and services trade and investment.

Evan Gershkovich (writer) has worked in Thailand as a freelance journalist covering forest conflict and as a communications specialist for a community forestry NGO.

Francis Wade (writer) is a freelance journalist based between Southeast Asia and London, with a focus on Myanmar. He worked as a journalist in Thailand for six years.

Greg Jorgensen (writer) writes about expat life in Thailand at GregToDiffer.com, and was co-creator and co-host of BangkokPodcast.com.

Ingo Puhl (contributing editor) is a co-founder of the South Pole Group, a World Economic Forum-recognized, clean-tech social enterprise, and an angel investor in companies that seek to create positive social and environmental impact at scale in the fields of food, consumer choice, consumer loyalty/reward and cause-related fundraising. He is also co-founder of Collaborative Designs, an investment company and accelerator that nourishes the execution of ideas that benefit communities and the planet.

Inhee Chung (writer) leads the Sustainability and Safeguards program at the Global Green Growth Institute (GGGI). Previously, she managed projects on biodiversity, cleaner production and green buildings at ERM and UNEP.

Jim Algie (editor and writer) is a Thailand veteran of Scottish-Canadian vintage who has authored a number of acclaimed books on the kingdom, and contributed a chapter to EDM's history book, Americans in Thailand.

Khan Ram-Indra (writer) is an environmental economist by training. He has worked on sustainable development, climate change, and clean energy for USAID, the British Embassy and private sector. He is currently serving as the Thailand Program Manager for the Global Green Growth Institute.

Kim Atkinson (editor) spent his career as an editor for UN agencies in Bangkok, Rome and West Africa, then returned to Bangkok to continue editing and teach writing.

Luxana Kiratibhongse (art director) graduated with a Fine Arts degree from Macalester College in the US. A graphic designer for more than 10 years, she has been commissioned by clients in Thailand and around the world for marketing, advertising and editorial projects.

Mark Fenn (writer) was a British journalist based in Thailand.

Mick Elmore (writer), a journalist of 30 years, has filed stories for nearly 100 publications from 20-plus countries. Based in Thailand since 1992, Mick earned his Southeast Asian Studies Masters at Chulalongkorn University and teaches there.

Molraudee Saratun (writer) is an Assistant Professor at the College of Management, Mahidol University. Her research focuses on human resource management, corporate sustainability and the Sufficiency Economy Philosophy.

Nelly Sangrujiveth (writer) has a JD and LLM in Environmental Law from the University of Oregon. Since graduating, she has worked with donor organization projects seeking to increase the implementation of clean energy in Asia through innovative financing and good governance.

Nichapat Chaokchaingamsangh (designer) earned her School of Fine and Applied Arts from Bangkok University.

Nicholas Grossman (editor-in-chief) has produced 10 books on Thailand, including *Thailand: 9 Days in the Kingdom*, *Chronicle of Thailand*, *Thailand At Random*, *Americans in Thailand*, *A History of the Thai-Chinese*, and *King Bhumibol Adulyadej: A Life's Work*.

Nikola Stalevski (writer) develops and promotes sustainability interventions in Bangkok for international cooperation agencies (GIZ and others). His passions are climate change mitigation, cities and transport.

Nina Wegner (editor and writer), is a freelance journalist who writes about indigenous issues and corporate responsibility in developing countries. Her work has been published in Al Jazeera, The Atlantic and The Huffington Post, among others.

Noel Boivin (writer), a Canadian writer and communications specialist, worked in media in Bangkok for a decade before joining the United Nations Education, Scientific and Cultural Organisation (UNESCO) as a media and communications officer.

Patima Klinsong (writer), a graduate in Technical Communication from Illinois Institute of Technology, has been a journalist, writer and translator for over 15 years.

Patinya Rojnukkarin (art director) earned her MFA from Minneapolis College of Art & Design in Visual Studies and BFA from Silpakorn University. She has 15 years of graphic and motion graphic experience.

Pattraporn Yamla-or (writer) is the co-founder of Sal Forest Co, Ltd., which furthers public discourse on sustainable business by conducting research on key sustainability issues as well as running workshops and events.

Phisanu Phromchanya (writer) has over 15 years of financial journalism experience with international wire services. He is now a consultant for an anti-corruption initiative in the private sector.

Purnama Pawa (assistant editor) is an editorial assistant, and sales and marketing manager for EDM. She holds a BA in Communications Management from Chulalongkorn University.

Raviprapa Srisartsanarat (writer) is an international development specialist with more than 17 years of experience in designing, managing, monitoring and advising on development programs for leading international donors, NGOs, and Thai government agencies.

Siree Simaraks (designer) has over six years of experience in graphic design. She holds a bachelor's degree in Graphic Design from the Faculty of Architecture, Urban Design and Creative Arts at Mahasarakham University.

Sofia Mitra-Thakur (writer) is a British journalist and writer, currently working at the Bangkok Post. She has previously worked for the South China Morning Post, The Telegraph, and The Independent, and Engineering & Technology Magazine.

Surasak Glahan (writer) has over 10 years in journalism and communications, including as a reporter at the Bangkok Post. He has worked for regional and international NGOs and inter-governmental organizations.

Surasak Tumcharoen (writer) was a Bangkok Post political news reporter for over 25 years. He is currently a correspondent at Xinhua News Agency.

Sutawan Chanprasert (assistant editor) had worked as a human rights researcher during the three years prior to joining EDM's editorial staff.

Tibor Krausz (writer) is a widely published writer and journalist. He is a lecturer at Bangkok University's International College and has worked as a consultant for the UN.

Tom Metcalfe (writer) is a journalist and filmmaker with a focus on science, environment and Asia-Pacific region.

Wasant Techawongtham (writer) was a former deputy news editor for *Bangkok Post*. Currently, he is a freelance writer and editor.

Will Baxter (writer and editor) is an American journalist and photographer who has been based in Southeast Asia since 2003. His work focuses mostly on human rights, conflict and development issues.

Woranuj Watts (writer) graduated in history from Thammasat University and worked as a business journalist for 15 years. She is now a freelance writer and translator.

This project was inspired by the launch of the

As parts of this book are derived from our publication, *Thailand's Sustainable Development Sourcebook*, we would like to acknowledge that project's sponsors:

We would also like to thank the project partners:

Advance Readers*

Ariya Arunin
M.R. Chakrarot Chitrabongs
Supachet Chansarn
William Klausner
Orathai Kokpol
Sucharit Koontanakulvong
Usa Lertsrisantad
Kiatanantha Lounkaew
Wimonthip Musikaphan
Willem Niemeijer
Deunden Nikomborirak
Sumeth Ongkittikul
Anand Panyarachun
Kanchana Patarachoke

Sompop Pattanariyankool
Prempreeda Pramoj Na Ayutthaya
Nipon Poapongsakorn
Banyong Pongpanich
Pranee Srihaban
Jittima Srisuknam
Seree Supharatid
Viroj Tancharoensathien
Somkiat Tangkitvanich
Ronnakorn Triraganon
Nalinee Thongtam
Supat Wangwongwatana
Doris Wibunsin
Ismail Wolff

*For the work that appeared in Thailand's Sustainable Development Sourcebook

Directory

Organizations

A

ActionAid
www.actionaid.org

ASEAN Cooperation on Environment
environment.asean.org

ASEAN Intergovernmental Commission on Human Rights (AICHR)
aichr.org

Asia Cooperation Dialogue (ACD)
www.acd-dialogue.org

Asian Development Bank
www.adb.org

Asian Disaster Preparedness Center (ADPC)
www.adpc.net

Asia Pacific Forum on Women, Law and Development (APWLD)
apwld.org

Asia Research Center for Migration (ARCM) – Institute of Asian Studies (IAS), Chulalongkorn University
www.arcmthailand.com

Association for the Promotion of the Status of Women
www.apsw-thailand.org

Association of Southeast Asian Nations (ASEAN)
asean.org

Asylum Access Thailand
asylumaccess.org

Australia-Asia Program to Combat Trafficking in Persons (AAPTIP)
www.aaptip.org

B

Bangchak Petroleum Public Co Ltd
www.bangchak.co.th

Biodegradable Packaging for Environment Co Ltd
www.thaibpe.com

Big Blue Conservation
www.bigblueconservation.com

Bird Conservation Society of Thailand (BCST)
www.bcst.or.th

C

Center for Oceanic Research and Education
coresea.com

Chiang Mai University's Forest Restoration Research Unit (FORRU-CMU)
www.forru.org

Climate Change Management and Coordination Division
web2.onep.go.th

Community-Based Tourism Institute (CBT-I)
www.cbt-i.org

Community Organization Development Institute (CODI)
www.codi.or.th

Constitutional Court of the Kingdom of Thailand
www.constitutionalcourt.or.th

Chaipattana Foundation
www.chaipat.or.th

Change.org
www.change.org

ChangeFusion
changefusion.org

Chulabhorn Research Institute
www.cri.or.th

Crown Property Bureau
www.crownproperty.or.th

D

Deep South Watch
www.deepsouthwatch.org

Department of Alternative Energy Development and Efficiency
www.dede.go.th

Department of Industrial Promotion under the Ministry of Industry
english.dip.go.th

Department of Labour Protection and Welfare
www.labour.go.th

Department of National Parks, Wildlife and Plant Conservation
portal.dnp.go.th

Department of Water Resources under the Ministry of Natural Resources and Environment (in Thai only)
www.dwr.go.th

Department of Women's Affairs and Family Development (DWF)
www.dwf.go.th

Distance Education Institute
www.dei.ac.th

Doi Chaang Coffee Original Company
www.doichaangcoffee.com

Doi Kham Food Products Co Ltd
www.doikham.co.th

Doi Tung Development Project
www.doitung.org

E

Electricity Generating Authority of Thailand (EGAT)
www.egat.co.th

EnerGaia
energaia.com

Energy Policy and Planning Office under Ministry of Energy
www.eppo.go.th

F

Faculty of Fisheries, Kasetsart University
www.fish.ku.ac.th

Federation of Thai Industries
www.fti.or.th

Food and Agricultural Organization of the United Nations (FAO)
www.fao.org

Foundation for Consumers
en.consumerthai.org

Foundation for Women (FFW)
www.womenthai.org

Foundation of Virtuous Youth
vyouth.org

Future Innovative Thailand Institute
fit.or.th

G

Gender and Development Research Institute (GDRI)
www.gdrif.org

Government Pension Fund
www.gpf.or.th

Global Alliance Against Trafficking Women (GAATW)
www.gaatw.org

Gracz Biodegradbale Packaging
www.gracz.co.th

Greenpeace
www.greenpeace.org

H

Habitat for Humanity Thailand
www.habitatthailand.org

Human Development Forum Foundation (HDFF)
hdff.org

Human Rights Lawyers Association (HRLA)
naksit.org

Hydro and Agro Informatics Institute (HAII) under the Ministry of Science and Technology
www.haii.or.th

I

Institute of Marine Science: Burapha University
www.buu.ac.th

Integrated Tribal Development Program
www.itdpinternational.org

International Organization for Migration (IOM): Thailand
th.iom.int

International Union for Conservation of Nature and Natural Resources (IUCN)
www.iucn.org

K

Kasetsart University
www.ku.ac.th

Kasetsart University Research and Development Institute (KURDI)
www2.rdi.ku.ac.th

Center for Agricultural Biotechnology (CAB)
www.cab.kps.ku.ac.th

Kasetsart Agricultural and Agro-Industrial Product Improvement Institute
webportal.ku.ac.th

Kenan Institute Asia
www.kenan-asia.org

King Prajadhipok's Institute (KPI)
kpi.ac.th

L

Law Reform Commission of Thailand (LRCT)
www.lrct.go.th

Living River Siam Association
www.livingriversiam.org

M

Mahidol University, College of Management
www.cmmu.mahidol.ac.th

Mae Fah Luang Foundation Under Royal Patronage
www.maefahluang.org

Mekong Institute (MI)
www.mekonginstitute.org

Mercy Center
www.mercycentre.org

Ministry of Agriculture and Cooperatives
eng.moac.go.th

Department of Agriculture
www.doa.go.th

Department of Agricultural Extension
www.doae.go.th

Ministry of Education
www.en.moe.go.th

Ministry of Energy
energy.go.th

Ministry of Foreign Affairs
www.mfa.go.th

DIRECTORY

Ministry of Industry
webnew.industry.go.th

Ministry of Labor
www.mol.go.th

Ministry of Public Health
eng.moph.go.th

Ministry of Social Development and Human Security
www.m-society.go.th

Ministry of Transport
www.mot.go.th

Mirror Foundation
www.mirror.or.th

N

National Innovation Agency
www.nia.or.th

National Institute of Development Administration
www.nida.ac.th

O

Office of the Election Commission of Thailand
www.ect.go.th

Office of the National Anti-Corruption Commission (NACC)
www.nacc.go.th

Office of the National Economic and Social Development Board
www.nesdb.go.th

Office of the National Human Rights Commission of Thailand (NHRC)
www.nhrc.or.th

Office of the Ombudsman Thailand
www.ombudsman.go.th

Office of the Royal Development Projects Board
www.rdpb.go.th

Office of SMEs Promotion (OSMEP)
www.sme.go.th

One Tambon, One Product (OTOP)
www.thaitambon.com

Organic Agriculture Certification Thailand (ACT)
www.actorganic-cert.or.th

Oxfam
www.oxfam.org.uk

P

Pid Thong Lang Phra Foundation (Royal Initiative Discovery Foundation)
www.pidthong.org

Plan International
plan-international.org

Population and Development Association (PDA)
www.pda.or.th

PTT Public Company Limited
www.pttplc.com

Pratthanadee Foundation
www.pratthanadee.org

Pranda Jewelry PCL
www.pranda.com

Pun Pun Center for Self Reliance
www.punpunthailand.org

Q

Queen Sirikit Institute
www.artsofthekingdom.com

R

Raithong Organics Farm
www.raitongorganicsfarm.com

Reef Biology Research Group, Department of Marine Science, Chulalongkorn University
www.marine.sc.chula.ac.th

Royal Irrigation Department under the Ministry of Agriculture and Cooperatives
www.rid.go.th

Royal Project Foundation (in Thai only)
www.royalprojectthailand.com

Right To Play
righttoplay.or.th

S

Sal Forest Co Ltd
www.salforest.com

Siam Cement Group (SCG)
www.scg.co.th

Small and Medium Enterprise Development Bank of Thailand (SME Bank)
www.smebank.co.th

Socialgiver.com
www.socialgiver.com

Solar Power Company Group (SPCG)
www.spcg.co.th

SUPPORT Arts and Crafts International Centre of Thailand (Public Organization)
www.sacict.net

Sustainability at King Mongkut's University of Technology Thonburi (KMUTT)
sustainable.kmutt.ac.th

T

Teach for Thailand (in Thai only)
www.teachforthailand.org

Thai Chamber of Commerce and Board of Trade of Thailand
www.thaichamber.org

Thai Health Promotion Foundation
en.thaihealth.or.th

Thai Industrial Standards Institute (TISI)
www.tisi.go.th

Thailand Automotive Institute (TAI)
www.thaiauto.or.th

Thailand Board of Investment (BOI)
www.boi.go.th

Thailand Business Coalition on AIDS (TBCA)
www.tbca.or.th

Thailand Development Research Institute (TDRI)
tdri.or.th

Thailand Environment Institute (TEI)
www.tei.or.th

Thailand International Cooperation Agency (TICA) under the Ministry of Foreign Affairs
www.tica.thaigov.net

Thailand Knowledge Park (TK Park)
www.tkpark.or.th

Thailand Sustainable Development Foundation
www.tsdf.or.th

Thailand Textile Institute (THTI)
www.thaitextile.org

Thailand Today
www.thailandtoday.in.th

Thai Lawyers for Human Rights
tlhr2014.wordpress.com

Thai National Influenza Center under the Department of Medical Sciences (DMSc)
www.thainihnic.org

Thai Red Cross Society
english.redcross.or.th

Tongsai Bay Hotel
www.tongsaibay.co.th

Tourism Authority of Thailand: 7greens
7greens.tourismthailand.org

U

UNESCO Bangkok
www.unescobkk.org

UNICEF: Thailand
www.unicef.org/thailand

United Nations Action for Cooperation Against Trafficking in Persons
un-act.org

United Nations Development Programme (UNDP) in Thailand
www.th.undp.org

United Nations Economic and Social Commission for Asia and the Pacific (UNESCAP)
www.unescap.org

Utokapat Foundation
www.utokapat.org

W

Wildlife Conservation Society (WCS)
thailand.wcs.org

Wildlife Fund Thailand
www.wildlifefund.or.th

Women's Health Advocacy Foundation
www.whaf.or.th

Women's Health and Reproductive Rights Foundation of Thailand
www.womenhealth.or.th

Wongpanit Recycling
www.wongpanit.com

World Health Organization (WHO)
www.who.int

Texts

#

"2016 World Health Statistics: Monitoring Health For the SDGs," World Health Organization (WHO)

"2016 World Press Freedom Index," Reporters Without Borders

A

"Achieving Skill Mobility in the ASEAN Economic Community: Challenges, Opportunities, and Policy Implications," Asian Development Bank, 2016

"A Critical Study of Thailand's Higher Education Reforms: The Culture of Borrowing" Rattana Lao, 2015

"Anti-Corruption Strategies: Understanding What Works, What Doesn't And Why?," United Nations Development Programme (UNDP), 2014

"ASEAN Annual Report 2015-2016," Association of Southeast Asia Nations (ASEAN)

"Asian Development Outlook 2016: Asia's Potential Growth," Asian Development Bank, 2016

B

"Building Resilience to Climate Change Impacts – Coastal Southeast Asia (BCR)," Angela Jöhl, International Union for Conservation of Nature, 2013

C

"Case Studies of Good Agricultural Practices (GAPs) of Farmers in Thailand," Khin Yadanar Oo, Center for Applied Economic Research, Kasetsart University, 2016

DIRECTORY

"Climate Change Impacts on Water Resources: Key Challenges to Thailand CC Adaptation," Sudtida Pliankarom Thanasupsin, Royal Irrigation Department

"Climate Vulnerability and Capacity Analysis Report: South of Thailand," CARE Desutschland-Luxemburg e.V. in cooperation with Raks Thai Foundation, 2013

"Commercial Bank Innovations in Small and Medium-Sized Enterprise Finance: Global Models and Implications for Thailand," Asian Development Bank, 2016

"Cost of Implementing National Biodiversity Strategies and Actions in Thailand: A Conceptual Approach," United Nations Development Programme (UNDP), 2016

D

"Drivers Affecting Forest Change in the Greater Mekong Subregion (GMS): An Overview," Food and Agricultural Organization of the United Nations (FAO), 2015

"Commercial Bank Innovations in Small and Medium-Sized Enterprise Finance: Global Models and Implications for Thailand," Asian Development Bank, 2016

E

"Education and Knowledge in Thailand: The Quality Controversy," Alain Mounier and Phasina Tangchuang, 2010

"Education for People and Planet: Creating Sustainable Futures for All," United Nations Educational, Scientific and Cultural Organization, 2016

"Employment Practices and Working Conditions in Thailand's Fishing Sector," ILO and Asian Research Center for Migration under the Institute of Asian Studies, Chulalongkorn University, 2013

Explaining Ocean Warming: Causes, Scales, Effects and Consequences, International Union for Conservation of Nature and Natural Resources (IUCN), 2016

F

"Farmers' Knowledge, Attitude and Practice Toward Organic Vegetables Cultivation in Northeast Thailand," Shimul Mondal, Theerachai Haitook, and Suchints Simaraks, Kasetsart Journal (Social Science) 35: 158-166, 2014

"Fatal Journeys: Tracking Lives Lost during Migration," International Organization for Migration (IOM), 2014

"Forest Landscape Restoration For Asia-Pacific forests," Food and Agricultural Organization of the United Nations (FAO), 2016

Forging Peace in Southeast Asia: Insurgencies, Peace Processes, and Reconciliation, Zachary Abuza, Rowman & Littlefield Publishers, 2016

"Freedom of the Press 2016," Freedom House

G

"Global Peace Index Report," Institute for Economics and Peace (IEP), 2016

"Global Status Report on Alcohol and Health," World Health Organization (WHO), 2014

"Greater Mekong Sub-region Statistics on Growth, Infrastructure, and Trade," Asian Development Bank, 2016

"Green Management Model for Eco-farm in Thailand," Suthet Chandrucka, Nuttiya Tantranont, and Manat Suwan, Universal Journal of Management, 2015

H

"Health in 2015: from MDGs to SDGs," World Health Organization, 2015

"Human Development Report 2015," United Nations Development Programme (UNDP)

L

"Living Planet Report," World Wildlife Fund (WWF), 2014

M

"Mainstreaming Climate Change into Community Development Strategies and Plans: A Case Study in Thailand," Suppakorn Chinvanno and Vichien Kerdsuk, Adaption Platform Knowledge, 2013

P

"Parks for Life: Why We Love Thailand's National Parks," Songtam Suksawang and Jeffrey A. McNeely, Department of National Parks, Wildlife, and Plant Conservation and United Nations Development Programme (UNDP), 2015

"Planning for Change: Guidelines for National Programmes on Sustainable Consumption and Production," United Nations Environment Programme (UNEP), 2008

Positioning the ASEAN Community in an Emerging Asia: Thai Perspectives, Department of ASEAN Affairs, Ministry of Foreign Affairs of the Kingdom of Thailand, 2016

R

Renewable Energy: Power for a Sustainable Future, Godfrey Boyle, Oxford University Press, 2012

S

"SDG Index & Dashboards - A Global Report" Jeffrey D. Sachs, Guido Schmidt-Traub, Christian Kroll, David Durand-Delacre, and Katerina Teksoz, Betelsmann Stiftung and Sustainable Development Solutions Network (SDSN), 2016

Six Degrees: Our Future on a Hotter Planet, Mark Lynas, National Geographic, 2008

"South-east Asia Migrant Crisis: Numbers Are Now 'Alarming', Talks Told," The Guardian, May 29, 2015

"Sufficiency Economy Philosophy and Development," Chaiwat Wibulswadi, Priyanut Dharmapiya, Kobsak Pootrakul, 2016

Sufficiency Thinking: Thailand's Gift to an Unsustainable World, Gayle C. Avery and Harald Bergsteiner, 2016

"Sustainable Intensification of Aquaculture in the Asia-Pacific Region," Food and Agriculture Organization of the United Nations (FAO), 2016.

"Sustainable Life Path Concept: Journeying toward Sustainable Consumption," Sue L.T. McGregor, Journal of Research for Consumes, 2013

"Sustainable Management of Community-Based Tourism in Thailand," Nopparat Satarat, National Institute of Development Administration (NIDA), 2010

T

Tearing Apart the Land: Islam and Legitimacy in Southern Thailand, Duncan McCargo, Cornell University Press, 2008

"Thailand Annual Report 2015," World Wide Fund for Nature (WWF)

"Thailand: Industrialization and Economic Catch-Up," Asian Development Bank, 2015

"Thailand's Progress in Agriculture: Transition and Sustained Productivity Growth," Henri Leturque and Steve Wiggins, The Overseas Development Institute (ODI), 2011

Thailand's Sustainable Development Sourcebook, Editions Didier Millet, 2015

The Age of Sustainable Development, Jeffrey D. Sachs, Columbia University Press, 2015

The Bottom Billion: Why the Poorest Countries are Falling and What Can Be Done About It, Paul Collier, Oxford University Press, 2008

"The Economics of Climate Change in Southeast Asia: A Regional Review," Asian Development Bank, 2009

The End of Poverty: Economic Possibilities for Our Time, Jeffrey D. Sachs, Penguin Books, 2006

"The Global Competitiveness Report 2016-2017," Klaus Schwab, World Economic Forum

"The Impact of Finance on the Performance of Thai Manufacturing Small and Medium-Sized Enterprises," Asian Development Bank, 2016

"The Kingdom of Thailand Health System Review," Pongpisut Jongudomsuk et al, World Health Organization on behalf of the Asia Pacific Observatory on Health Systems and Policies, Vol. 5 No.5 2015

The Politics of Uneven Development: Thailand's Economic Growth in Comparative Perspective, Richard F. Doner, Cambridge University Press, 2009

The Price of Inequality: How Today's Divided Society Endangers Our Future by Joseph Stiglitz, W.W. Norton & Company, 2013

The Life You Can Save: How to Do Your Part to End World Poverty by Peter Singer, Random House Trade Paperbacks, 2010

"The State of Food and Agriculture: Social protection and Agriculture: Breaking the Cycle of Rural Poverty," Food and Agriculture Organization of the United Nation (FAO), 2015

"The State of Food Security in the World," Food and Agriculture Organization of the United Nation (FAO), 2015

"The United Nations World Water Development Report 2016: Water and Jobs," United Nations Educational, Scientific and Cultural Organization, 2016

This Changes Everything: Capitalism vs Climate Change, Naomi Klein, Simon & Schuster, 2014

Truth on Trial in Thailand: Defamation, Treason, and lèse-majesté by David Streckfuss, Routledge, 2011

"Twenty Years of Failure," Greenpeace, 2015

U

Unequal Thailand: Aspects of Income, Wealth and Power, Pasuk Phongpaichit and Chris Baker, 2015

"UN Women Annual Report 2015-2016," United Nations Entity for Gender Equality and the Empowerment of Women

W

"World Health Statistics 2016," World Health Organization (WHO)

"World Migration Report 2015," Migrants and Cities: New Partnerships to Manage Mobility, International Organization for Migration (IOM), 2015

Picture Credits

Every effort has been made to trace copyright holders of images in this book. In the event of error or omissions, appropriate credit will be made in future editions of *A Call to Action: Thailand and the Sustainable Development Goals*.

Athit Perawongmetha: 109
Bangkok Post: 32, 58
Ben Simmons: 117
Big Trees Project: 121, 154
Bohnchang Koo: 97
Charoon Thongnual: 35, 46
Chawalit Poompo/Rice Department: 38
Chien-Chi Chang/Magnum Photos: 8
Community Organization Development Institute: 119
Crown Property Bureau: 40, 56, 145
Doi Tung Development Project: 36
FORRU-CMU: 150
Getty Images: 16, 21, 34, 37, 51, 53, 66, 72, 78, 80, 88, 92, 96, 98, 106, 110, 111, 113, 132, 133, 135, 143, 152, 160, 164, 167, 168, 170, 174
Greenpeace: 86
Huai Hong Khrai Royal Development Study Center: 43, 156
Jinnawat Pumpoung: 24
Nat Sumanatemeya: 146
Oxfam: 177
Plan Creations Company Limited: 128 all
Prasarn Sangpaitoon/Rambhai Barni Rajabhat University: 148
Raitong Organics: 45
SCG: 23, 100, 130
Shutterstock: 5, 42, 55, 76, 79, 81, 84, 114, 116, 123, 124, 127, 155
Sufficiency School Center: 60, 63, 65
SUPPORT Foundation: 73
Thailand International Cooperation Agency: 172, 175
The Royal Discovery Initiative Foundation: 3
The Thai Silk Company: 129
Wildlife Conservation Society: 158
Will Baxter: 30, 33, 59, 71, 94
World Wildlife Fund: 159
Yann Arthus-Bertrand: 136 (13°59′ N, 100°25′ E), 140 (8°20′ N, 98°30′ E)

"The happiness and prosperity that people seek can be achieved, but by actions that are ethical in intention and execution, not by chance or by fighting and grabbing from others. True prosperity is something creative because it gives benefit to others and to people in general as well."

– His Majesty King Bhumibol Adulyadej